SAXON MATH™ 2

Student Workbook

Part 1

Nancy Larson

with Roseann Paolino
Maureen Hannan

SAXON™

A Harcourt Achieve Imprint

www.SaxonPublishers.com
1-800-284-7019

ISBN-13: 978-1-6003-2721-6 (Set)
ISBN-10: 1-6003-2721-4 (Set)
ISBN-13: 978-1-6003-2574-8 (Part 1)
ISBN-10: 1-6003-2574-2 (Part 1)
ISBN-13: 978-1-6003-2577-9 (Part 2)
ISBN-10: 1-6003-2577-7 (Part 2)

Printed in the United States of America

4 5 6 7 8 9 054 14 13 12 11 10 09 08

To the Teacher

This individual student unit contains all the necessary material for one student for the entire year.

The pages immediately following contain an assortment of individual recording forms developed for your convenience. Record scores from the Fact and Written Assessments on the front of each form and scores from the Oral Assessments on the back. Optional classroom versions of these forms are provided in the Monitoring Student Progress binder.

The student materials follow the recording forms in the order they will be used and include the following: Class Fact Practices, Fact Homeworks, and Fact Assessments; Lesson Worksheets; Guided Class Practices and Homeworks; Problem-Solving and Performance Task Worksheets; and Written Assessments. A Parent Letter follows the Guided Class Practice/Homework in Lesson 2 and should be sent home after this lesson. (For your convenience, a Spanish translation of this letter is provided in the Home Connections section of the Monitoring Student Progress binder and the Resources and Planner CD.) Also provided are multiple copies of the Meeting Worksheet, which students use daily during The Meeting.

These student units are designed to supplement the classroom kits. We suggest that you file the recording forms in the Monitoring Student Progress binder and integrate the remaining student materials into your file folders.

Fact and Written Assessments Individual Recording Form

Saxon Math 2 (Lessons 10, 15, 20, 25)

Student ______________________________

Teacher ________________ School Year ________

Fact Assessment 1		Written Assessment 1	Circle missed items.		Date ______
Lesson 10 ____ Score	**Doubles** Missed Facts:	Lesson 10 ____ Score	**1.** reads graph; identifies most	**2.** continues shape pattern	**3.** identifies one more, one less than a number
			4. writes time to the hour	**5.** identifies missing numbers to 30	**6.** addition facts: doubles
Fact Assessment 2		**Written Assessment 2**	**Circle missed items.**		Date ______
Lesson 15 ____ Score	**Adding 1 and 2; Doubles** Missed Facts:	Lesson 15 ____ Score	**1.** identifies ordinal position: 1st, 2nd, 4th, 5th; identifies middle	**2.** compares two numbers	**3.** number pattern: counts by 10's to 100
			4. identifies missing numbers to 40 on a hundred number chart	**5.** identifies right/left; draws circle, square	**6.** addition facts: +1
Fact Assessment 3		**Written Assessment 3**	**Circle missed items.**		Date ______
Lesson 20 ____ Score	**Adding 2; Review Facts** Missed Facts:	Lesson 20 ____ Score	**1.** identifies stories: SSM, SWA	**2.** continues shape pattern	**3.** writes two-digit numbers
			4. identifies equal parts of a whole	**5.** identifies one more, one less than a number	**6.** addition facts: doubles, +1, +0
Fact Assessment 4		**Written Assessment 4**	**Circle missed items.**		Date ______
Lesson 25 ____ Score	**Adding 9** Missed Facts:	Lesson 25 ____ Score	**1.** reads graph	**2.** identifies weekdays	**3.** identifies ordinal position to 12th
			4. identifies even numbers to 20	**5.** numbers clockface; writes, shows elapsed time to the hour	**6.** identifies shapes

Student ______________________________

Teacher ______________ School Year ________

Oral Assessment
Individual Recording Form
***Saxon** Math 2*

LESSON 10-2: Oral Assessment 1

Date __________

Identifying and Writing Numbers to 100

Materials: Math Folder, handwriting paper, pencil

• Show the child the hundred number chart on the Math Folder. ***"Point to the number 17."*** ***"... 82."*** ***"... to the right of 65."*** ***"... to the left of 23."***	• The following can be done as a class activity: ***"Write these numbers: 32 ... 71 ... 80 ... 16 ... 53 ... 12 ... 19."***

Comments: ______________________________

LESSON 20-2: Oral Assessment 2

Date __________

Identifying Ordinal Position

Materials: pattern blocks

• Place 12 pattern blocks in a row. ***"Point to the first block."*** ***"... fourth block."*** ***"... eighth block."*** ***"... eleventh block."***

Comments: ______________________________

Fact and Written Assessments Individual Recording Form

Saxon Math 2 (Lessons 30, 35, 40, 45)

Student ______________________

Teacher ______________ School Year ______________

Fact Assessment 5		Written Assessment 5	Circle missed items.		Date ______
Lesson 30 ____ Score	**Doubles + 1** Missed Facts:	Lesson 30 ____ Score	**1.** draws SSM story; writes number sentence; solves	**2.** identifies temperature to 10°	**3.** shows one half of a whole
			4. compares three two-digit numbers	**5.** identifies fractions: $\frac{1}{2}$, $\frac{1}{3}$, $\frac{1}{6}$	**6.** addition facts: +2, +9
Fact Assessment 6		**Written Assessment 6**	**Circle missed items.**		**Date ______**
Lesson 35 ____ Score	**Sums of 8 and 9; Review Facts** Missed Facts:	Lesson 35 ____ Score	**1.** draws SSM story; writes number sentence; solves	**2.** shows one half of a square in two ways	**3.** counts money (dimes)
			4. identifies ordinal position to 6th; days of the week; months	**5.** identifies odd numbers to 19	**6.** addition facts: doubles, +0, +1, +2, +9, doubles + 1
Fact Assessment 7		**Written Assessment 7**	**Circle missed items.**		**Date ______**
Lesson 40 ____ Score	**Sums of 10; Review Facts** Missed Facts:	Lesson 40 ____ Score	**1.** draws SWA story; writes number sentence; solves	**2.** numbers clockface; writes, shows time to the half hour	**3.** draws tally marks
			4. counts money (dimes, pennies)	**5.** reads graph	**6.** writes addition/ subtraction fact family
Fact Assessment 8		**Written Assessment 8**	**Circle missed items.**		**Date ______**
Lesson 45 ____ Score	**Sums of 11; Review Facts** Missed Facts:	Lesson 45 ____ Score	**1.** draws SSM story; writes number sentence; solves	**2.** adds dimes, pennies; adds tens, ones	**3.** number patterns: counts by 10's, counts by 5's
			4. identifies fractions: halves, fourths, eighths	**5.** adds 10 to a multiple of 10	**6.** identifies missing addends: sums of 10

Student ______________________________

Teacher ________________ School Year ________

Oral Assessment
Individual Recording Form
***Saxon** Math 2*

LESSON 30-2: Oral Assessment 3

Date __________

Identifying Attributes of Shapes

Materials:
Attribute Shape pieces (from Lesson 21)

• Show the child two pieces that differ by both color and size.

"How are these pieces the same?"	*"How are these pieces different?"*

Comments: ________________________________

LESSON 40-2: Oral Assessment 4

Date __________

Counting by 10's and 5's

Materials:
none

"Count by 10's to 100."	*"Count by 5's to 50."*

Comments: ________________________________

Fact and Written Assessments Individual Recording Form

Saxon Math 2 (Lessons 50, 55, 60, 65)

Student ______________________

Teacher ______________ School Year ________

Fact Assessment 9		Written Assessment 9	Circle missed items.		Date ______
Lesson 50 ____ Score	**Sums of 12; Review Facts** Missed Facts:	Lesson 50 ____ Score	**1.** draws SWA story with extra information; writes number sentence; solves **4.** identifies, counts pairs	**2.** identifies horizontal/ vertical/oblique **5.** addition facts: +2, sums of 10, doubles + 1	**3.** adds dimes, pennies; shows amount using fewest dimes, pennies **6.** identifies missing numbers on a hundred number chart
Fact Assessment 10		**Written Assessment 10**	**Circle missed items.**		Date ______
Lesson 55 ____ Score	**Sums of 13 and 14; Review Facts** Missed Facts:	Lesson 55 ____ Score	**1.** SSM story (+10): writes number sentence; solves **4.** measures line segment using inches	**2.** logic problem; ordinal position to 5th **5.** number patterns: even numbers, odd numbers	**3.** identifies temperature to nearest 10° **6.** adds 10 to a two-digit number
Fact Assessment 11		**Written Assessment 11**	**Circle missed items.**		Date ______
Lesson 60 ____ Score	**Sums of 15, 16, 17, and 18; Review Facts** Missed Facts:	Lesson 60 ____ Score	**1.** SSM story (+10): writes number sentence; solves **4.** colors fractional parts of a whole: , ,	**2.** counts, draws tally marks **5.** compares, orders three two-digit numbers	**3.** counts money (nickels) **6.** writes dates using digits, words
Fact Assessment 12		**Written Assessment 12**	**Circle missed items.**		Date ______
Lesson 65 ____ Score	**100 Addition Facts** Missed Facts:	Lesson 65 ____ Score	**1.** SSM story: writes number sentence; solves **4.** draws pairs, counts all	**2.** numbers, identifies points on number line **5.** draws lines of symmetry	**3.** counts money (dimes, nickels, pennies) **6.** adds 10; adds without regrouping

M2(3e)-IRF-3a

Student ______________________ Teacher ______________ School Year ________

Oral Assessment
Individual Recording Form
Saxon Math 2

LESSON 50-2: Oral Assessment 5

Date ________

Materials:
bag of fraction pieces (from Lesson 34)

Identifying Halves, Fourths, and Eighths

A. ***"Which piece is one half of the yellow circle?"***
B. ***"Which piece is one fourth of the yellow circle?"***
C. ***"Which piece is one eighth of the yellow circle?"***
D. ***"Cover the yellow circle using eighths."***
E. ***"How many eighths did you use?"***
F. ***"Cover the yellow circle using fourths."***
G. ***"How many fourths did you use?"***

A	B	C	D	E	F	G

Comments: ________________________________

LESSON 60-2: Oral Assessment 6

Date ________

Materials:
none

Making Up Addition and Subtraction Stories

"Make up a some, some more story."	*"Make up a some, some went away story."*

- Reassess each child on questions answered incorrectly (or not answered) on Assessments 1–5.

Comments: ________________________________

Student ____________________

Teacher ____________ School Year ____________

Fact and Written Assessments
Individual Recording Form
***Saxon** Math 2 (Lessons 70, 75, 80, 85)*

Fact Assessment 13-1, 13-2		Written Assessment 13		Circle missed items.	Date ______
Lesson 70 ____ Score	**Subtracting 1 and 0** Missed Facts:	Lesson 70	**1.** SSM story (money): writes number sentence; solves; writes as money	**2.** orders five two-digit numbers	**3.** writes fraction of a whole: $\frac{1}{2}$, $\frac{3}{4}$
____ Score	**100 Addition Facts** Missed Facts:	____ Score	**4.** draws line segment using inches	**5.** identifies, counts angles	**6.** adds without regrouping; adds four and five one-digit numbers
Fact Assessment 14-1, 14-2		**Written Assessment 14**		**Circle missed items.**	Date ______
Lesson 75 ____ Score	**Subtracting 2, 1, and 0** Missed Facts:	Lesson 75	**1.** SSM story (money): writes as money; writes number sentence; solves	**2.** draws dozen; identifies half dozen	**3.** writes, shows time to the half hour; identifies a.m./p.m.
____ Score	**100 Addition Facts** Missed Facts:	____ Score	**4.** draws, counts money (dimes, nickels, pennies)	**5.** identifies congruent shapes	**6.** adds with regrouping
Fact Assessment 15-1, 15-2		**Written Assessment 15**		**Circle missed items.**	Date ______
Lesson 80 ____ Score	**Subtracting 3 and 2** Missed Facts:	Lesson 80	**1.** SWA story (dozen): writes number sentence; solves	**2.** reads Venn diagram	**3.** identifies temperature to 2°
____ Score	**100 Addition Facts** Missed Facts:	____ Score	**4.** writes fraction for shaded part of a whole: $\frac{1}{6}$, $\frac{3}{4}$	**5.** draws line segments using half inches	**6.** adds with regrouping
Fact Assessment 16-1, 16-2		**Written Assessment 16**		**Circle missed items.**	Date ______
Lesson 85 ____ Score	**Subtracting 4 and 3** Missed Facts:	Lesson 85	**1.** SSM story with regrouping: writes number sentence; solves	**2.** shows temperature to 2°	**3.** orders five two- or three-digit numbers
____ Score	**Subtracting 0–4** Missed Facts:	____ Score	**4.** writes, shows time to five minutes	**5.** addition/subtraction: +10, –10, –1; adds three one-digit numbers	**6.** adds with regrouping

Student ______________________

Teacher ______________ School Year ________

Oral Assessment
Individual Recording Form
Saxon *Math 2*

LESSON 70-2: Oral Assessment 7

Date ________

Making Congruent Shapes; Dividing a Shape in Half

Materials:
Teacher Master 70-2
geoboards
(1 per child)
geobands
(5 per child)

This assessment can be done with 4–6 children at a time. Separate the children for the assessment.

• Show the child a copy of Teacher Master 70-2. • Give the child a geoboard and some geobands. ***"Copy this shape on your geoboard."***	• Make the following rectangle on the child's geoboard: • Give the child another geoband. ***"Use this band to divide the rectangle in half."*** ***"Can you divide it in half a different way?"***

Comments: ______________________

LESSON 80-2: Oral Assessment 8

Date ________

Making and Reading a Graph

Materials:
baskets of color tiles
of mixed colors

This assessment can be done with 4–6 children at a time. Separate the children for the assessment.

"Take a handful of color tiles."
A. ***"Make a graph to show the colors of the tiles you picked."***
• Allow time for the child to make the graph.
B. ***"Tell me about your graph."***
C. ***"How many more (color) tiles are there than (color) tiles?"***

A. graphs with one-to-one correspondence	B. describes the graph	C. compares columns

Comments: ______________________

Fact and Written Assessments Individual Recording Form

Saxon *Math 2 (Lessons 90, 95, 100, 105)*

Student ______________________________

Teacher ____________________ School Year ____________

Fact Assessment 17-1, 17-2		Written Assessment 17		Circle missed items.	Date ________
Lesson 90 ____ Score	**Subtracting 5 and 4** Missed Facts:	Lesson 90 ____ Score	**1.** SSM story with three addends: writes number sentence; solves	**2.** reads chart; identifies most/fewest; orders four three-digit numbers	**3.** measures line segments using half inches
____ Score	**Subtracting 0–5** Missed Facts:		**4.** identifies time to five minutes; identifies a.m./p.m.	**5.** writes number for picture of base ten blocks	**6.** subtracts 10; adds two and three two-digit numbers

Fact Assessment 18-1, 18-2		Written Assessment 18		Circle missed items.	Date ________
Lesson 95 ____ Score	**Subtracting 6 and 5** Missed Facts:	Lesson 95 ____ Score	**1.** SWA story (−10): writes number sentence; solves	**2.** draws pictograph with scale of 2	**3.** uses comparison symbols: >, <, =
____ Score	**Subtracting 0–6** Missed Facts:		**4.** draws, counts money (dimes, nickels, pennies); writes amount two ways	**5.** draws picture for number; writes number in expanded form	**6.** adds two and three two-digit numbers

Fact Assessment 19-1, 19-2		Written Assessment 19		Circle missed items.	Date ________
Lesson 100 ____ Score	**Subtracting 7 and 6** Missed Facts:	Lesson 100 ____ Score	**1.** SSM story with regrouping (money): writes as money; writes number sentence; solves	**2.** logic problem	**3.** shows time to five minutes
____ Score	**Subtracting 0–7** Missed Facts:		**4.** writes a fraction to show part of a set	**5.** reads, draws pictograph with scale of 2	**6.** subtracts two-digit numbers with and without regrouping

Fact Assessment 20-1, 20-2		Written Assessment 20		Circle missed items.	Date ________
Lesson 105 ____ Score	**Subtracting 8 and 7** Missed Facts:	Lesson 105 ____ Score	**1.** SWA story with regrouping: writes number sentence; solves	**2.** draws picture for number; writes number in expanded form; writes numbers using digits	**3.** draws, counts money (quarters)
____ Score	**Subtracting 0–8** Missed Facts:		**4.** uses comparison symbols: >, <, =	**5.** multiplies by 10	**6.** subtracts two-digit numbers with regrouping

Student ______________________

Teacher ______________ School Year ________

Oral Assessment Individual Recording Form
Saxon Math 2

LESSON 90-2: Oral Assessment 9

Date ________

Reading a Thermometer to the Nearest 2°

Materials:
thermometer

• Show the child the thermometer. ***"What is the temperature?"*** • Accept answers to the nearest 2°.

Comments: ______________________

LESSON 100-2: Oral Assessment 10

Date ________

Modeling and Describing Addition With Regrouping

Materials:
10 dimes
20 pennies
scrap paper

• Write 36¢ + 47¢ on a piece of scrap paper. • Hand the child the coins. ***"Find this answer using these coins."***		• Write 36¢ + 47¢ on a piece of scrap paper. ***"Show how to find the answer for this problem."*** ***"Explain each step."***		
shows money amounts using coins	names the sum	sets up problem	adds correctly	explains steps (references money or place value)

Comments: ______________________

Fact and Written Assessments Individual Recording Form

Saxon Math 2 (Lessons 110, 115, 120, 125)

Student ______________________

Teacher ______________ School Year __________

Fact Assessment 21-1, 21-2		Written Assessment 21		Circle missed items.	Date ______
Lesson 110 _____ Score	**Subtracting 9 and 8** Missed Facts:	Lesson 110 _____ Score	**1.** SSM story with extra information: writes number sentence; solves	**2.** shows one half of even number by sharing	**3.** rounds to nearest 10
_____ Score	**100 Subtraction Facts** Missed Facts:		**4.** identifies geometric solids: cone, sphere, cylinder, cube	**5.** multiplies by 1, 10, 100	**6.** adds/subtracts two-digit numbers with regrouping
Fact Assessment 22-1, 22-2		**Written Assessment 22**		**Circle missed items.**	**Date ______**
Lesson 115 _____ Score	**Multiplying by 5** Missed Facts:	Lesson 115 _____ Score	**1.** SWA story with extra information: writes number sentence; solves	**2.** shows one half of even, odd numbers by sharing	**3.** counts money (quarters, dimes, nickels); writes amount two ways
_____ Score	**100 Subtraction Facts** Missed Facts:		**4.** draws, identifies parallel lines	**5.** writes, shows, identifies time to the minute	**6.** adds/subtracts two-digit numbers with regrouping
Fact Assessment 23-1, 23-2		**Written Assessment 23**		**Circle missed items.**	**Date ______**
Lesson 120 _____ Score	**Multiplying by 2** Missed Facts:	Lesson 120 _____ Score	**1.** draws equal groups story; writes number sentence; solves	**2.** measures using centimeters; finds perimeter	**3.** draws line segment using centimeters
_____ Score	**100 Subtraction Facts** Missed Facts:		**4.** draws, counts money (quarters, dimes, nickels); writes amount two ways	**5.** draws bar graph with scale of 2	**6.** multiplies by 5; adds/subtracts two- and three-digit numbers with regrouping
Fact Assessment 24-1, 24-2		**Written Assessment 24**		**Circle missed items.**	**Date ______**
Lesson 125 _____ Score	**Multiplying by 3** Missed Facts:	Lesson 125 _____ Score	**1.** SSM story with three-digit numbers: writes number sentence; solves	**2.** reads, draws bar graph with scale of 2	**3.** identifies right angles
_____ Score	**100 Subtraction Facts** Missed Facts:		**4.** shows mixed number	**5.** number patterns: counts by 5's, counts by 10's	**6.** adds three-digit numbers; subtracts two-digit numbers

Student ______________________

Teacher ______________ School Year ________

Oral Assessment
Individual Recording Form
***Saxon** Math 2*

LESSON 110-2: Oral Assessment 11

Date ________

Reading and Showing Time to Five-Minute Intervals

Materials:
demonstration clock
student clock

• Set the demonstration clock to show a time on a 5-minute mark. ***"It's morning." "What time is it?"***	• Give the child a student clock. ***"Show (5:45) on your clock."*** (Vary the time used.)

Comments: ______________________

LESSON 120-2: Oral Assessment 12

Date ________

Modeling and Describing Subtraction With Renaming

Materials:
cup of 10 dimes
cup of 20 pennies
scrap paper

• Give the child a cup of dimes and a cup of pennies. ***"Put 52¢ on the paper." "Give me 15¢ of that money." "How much money do you have left?"***				• Write *52¢ – 15¢* on a piece of scrap paper. ***"Show how to find the answer for this problem." "Explain each step."***		
recognizes that trading is necessary	trades a dime for 10 pennies	gives away 15¢	counts money	trades correctly	subtracts correctly	explains steps

Comments: ______________________

Fact and Written Assessments Individual Recording Form

***Saxon** Math 2 (Lessons 130, 135)*

Student ______________________

Teacher ______________ School Year ______________

Fact Assessment 25-1, 25-2		Written Assessment 25	Circle missed items.		Date ______
Lesson 130 ______ Score	**Multiplying by 4** Missed Facts:	Lesson 130 ______ Score	**1.** draws equal groups story; writes number sentence; solves	**2.** labels, writes number sentence for array	**3.** draws, identifies perpendicular lines
______ Score	**100 Subtraction Facts** Missed Facts:		**4.** identifies most likely; justifies answer	**5.** writes mixed number for picture	**6.** adds/subtracts three-digit numbers, money amounts

Fact Assessment 26		Written Assessment 26	Circle missed items.		Date ______
Lesson 135 ______ Score	**Multiplying by 2, 3, 4, and 5** Missed Facts:	Lesson 135 ______ Score	**1.** SSM story with three-digit numbers (money): writes number sentence; solves	**2.** reads, draws pictograph with scale of 2	**3.** identifies, places points on coordinate plane
			4. identifies time to the quarter hour	**5.** identifies best number sentence for estimating a sum	**6.** uses comparison symbols: >, <, =; multiplies by 2, 3, 4; divides by 2

Student ______________________

Teacher ______________ School Year ______

Oral Assessment
Individual Recording Form
***Saxon** Math 2*

LESSON 130-2: Oral Assessment 13

Date ________

Counting and Showing Money Amounts to $1.00

Materials:
5 quarters
10 dimes
10 nickels
20 pennies

This assessment can be done with 4–6 children at a time. Separate the children for the assessment.

• Hand the child a group of coins with a total value less than $1.00. Include at least one of each coin. (Give each child a different number of coins.) ***"Count the money."***	***"Show (82¢)."*** (Vary the money amounts used.) ***"Show (82¢) using different coins."***

• Reassess each child on questions answered incorrectly (or not answered) on Assessments 1–12.

Comments: ______________________

Name ____________________

Meeting Worksheet

***Saxon** Math 2 (first used in **Lesson 18**)*

Lesson ________ Time [:] Money ________ Secret Number ________

Lesson ________ Time [:] Money ________ Secret Number ________

Lesson ________ Time [:] Money ________ Secret Number ________

Lesson ________ Time [:] Money ________ Secret Number ________

Name ____________________

Meeting Worksheet

Saxon *Math 2 (first used in* ***Lesson 18****)*

Lesson ______ Time [:] Money ______ Secret Number ______

Lesson ______ Time [:] Money ______ Secret Number ______

Lesson ______ Time [:] Money ______ Secret Number ______

Lesson ______ Time [:] Money ______ Secret Number ______

Meeting Worksheet

***Saxon** Math 2 (first used in **Lesson 18**)*

Name ________________________________

Lesson ______ Time [:] Money ______ Secret Number ______

Lesson ______ Time [:] Money ______ Secret Number ______

Lesson ______ Time [:] Money ______ Secret Number ______

Lesson ______ Time [:] Money ______ Secret Number ______

Name ______________________________

Meeting Worksheet

***Saxon** Math 2 (first used in **Lesson 18**)*

Lesson ______ Time [:] Money ______ Secret Number ______

Lesson ______ Time [:] Money ______ Secret Number ______

Lesson ______ Time [:] Money ______ Secret Number ______

Lesson ______ Time [:] Money ______ Secret Number ______

Name ______________________________

Meeting Worksheet

***Saxon** Math 2 (first used in **Lesson 18**)*

Lesson ________ Time [:] Money ________ Secret Number ________

Lesson ________ Time [:] Money ________ Secret Number ________

Lesson ________ Time [:] Money ________ Secret Number ________

Lesson ________ Time [:] Money ________ Secret Number ________

Name ____________________

Meeting Worksheet

***Saxon** Math 2 (first used in **Lesson 18**)*

Lesson ______ Time [:] Money ______ Secret Number ______

Lesson ______ Time [:] Money ______ Secret Number ______

Lesson ______ Time [:] Money ______ Secret Number ______

Lesson ______ Time [:] Money ______ Secret Number ______

Name ______________________________

Meeting Worksheet

Saxon *Math 2 (first used in* ***Lesson 18****)*

Lesson ______ Time [:] Money ______ Secret Number ______

Lesson ______ Time [:] Money ______ Secret Number ______

Lesson ______ Time [:] Money ______ Secret Number ______

Lesson ______ Time [:] Money ______ Secret Number ______

Name ______________________

Meeting Worksheet

***Saxon** Math 2 (first used in **Lesson 18**)*

Lesson ______ Time [:] Money ______ Secret Number ______

Lesson ______ Time [:] Money ______ Secret Number ______

Lesson ______ Time [:] Money ______ Secret Number ______

Lesson ______ Time [:] Money ______ Secret Number ______

Name ______________________

Meeting Worksheet

***Saxon** Math 2 (first used in **Lesson 18**)*

Lesson ______ Time [:] Money ______ Secret Number ______

Lesson ______ Time [:] Money ______ Secret Number ______

Lesson ______ Time [:] Money ______ Secret Number ______

Lesson ______ Time [:] Money ______ Secret Number ______

Name ______________________

Meeting Worksheet

Saxon *Math 2 (first used in* ***Lesson 18****)*

Lesson ______ Time [:] Money ______ Secret Number ______

Lesson ______ Time [:] Money ______ Secret Number ______

Lesson ______ Time [:] Money ______ Secret Number ______

Lesson ______ Time [:] Money ______ Secret Number ______

Name ______________________________

Meeting Worksheet

Saxon *Math 2 (first used in* ***Lesson 18****)*

Lesson ________ Time [:] Money ________ Secret Number ________

Lesson ________ Time [:] Money ________ Secret Number ________

Lesson ________ Time [:] Money ________ Secret Number ________

Lesson ________ Time [:] Money ________ Secret Number ________

Name ______________________

Meeting Worksheet

***Saxon** Math 2 (first used in **Lesson 18**)*

Lesson ______ Time [:] Money ______ Secret Number ______

Lesson ______ Time [:] Money ______ Secret Number ______

Lesson ______ Time [:] Money ______ Secret Number ______

Lesson ______ Time [:] Money ______ Secret Number ______

Name ______________________

Meeting Worksheet

***Saxon** Math 2 (first used in **Lesson 18**)*

Lesson ________ Time [:] Money ________ Secret Number ________

Lesson ________ Time [:] Money ________ Secret Number ________

Lesson ________ Time [:] Money ________ Secret Number ________

Lesson ________ Time [:] Money ________ Secret Number ________

Name ______________________________

Meeting Worksheet

Saxon Math 2 (first used in Lesson 18)

Lesson ______ Time [:] Money ______ Secret Number ______

Lesson ______ Time [:] Money ______ Secret Number ______

Lesson ______ Time [:] Money ______ Secret Number ______

Lesson ______ Time [:] Money ______ Secret Number ______

Name ______________________

Meeting Worksheet

***Saxon** Math 2 (first used in **Lesson 18**)*

Lesson ________ Time [:] Money ________ Secret Number ________

Lesson ________ Time [:] Money ________ Secret Number ________

Lesson ________ Time [:] Money ________ Secret Number ________

Lesson ________ Time [:] Money ________ Secret Number ________

Meeting Worksheet

Saxon Math 2 (*first used in* **Lesson 18**)

Name ______________________

Lesson _____ Time [:] Money _____ Secret Number _____

Lesson _____ Time [:] Money _____ Secret Number _____

Lesson _____ Time [:] Money _____ Secret Number _____

Lesson _____ Time [:] Money _____ Secret Number _____

Name ______________________________

Meeting Worksheet

***Saxon** Math 2 (first used in **Lesson 18**)*

Lesson ________ Time [:] Money ________ Secret Number ________

Lesson ________ Time [:] Money ________ Secret Number ________

Lesson ________ Time [:] Money ________ Secret Number ________

Lesson ________ Time [:] Money ________ Secret Number ________

Name ______________________

Meeting Worksheet

***Saxon** Math 2 (first used in **Lesson 18**)*

Lesson ______ Time [:] Money ______ Secret Number ______

Lesson ______ Time [:] Money ______ Secret Number ______

Lesson ______ Time [:] Money ______ Secret Number ______

Lesson ______ Time [:] Money ______ Secret Number ______

Name ______________________

Meeting Worksheet

***Saxon** Math 2 (first used in **Lesson 18**)*

Lesson ______ Time [:] Money ______ Secret Number ______

Lesson ______ Time [:] Money ______ Secret Number ______

Lesson ______ Time [:] Money ______ Secret Number ______

Lesson ______ Time [:] Money ______ Secret Number ______

Meeting Worksheet

Saxon *Math 2 (first used in* ***Lesson 18****)*

Name ____________________

Lesson ______ Time [:] Money ______ Secret Number ______

Lesson ______ Time [:] Money ______ Secret Number ______

Lesson ______ Time [:] Money ______ Secret Number ______

Lesson ______ Time [:] Money ______ Secret Number ______

Name ______________________

Meeting Worksheet

***Saxon** Math 2 (first used in **Lesson 18**)*

Lesson ______ Time [:] Money ______ Secret Number ______

Lesson ______ Time [:] Money ______ Secret Number ______

Lesson ______ Time [:] Money ______ Secret Number ______

Lesson ______ Time [:] Money ______ Secret Number ______

Name ______________________________

Meeting Worksheet

***Saxon** Math 2 (first used in **Lesson 18**)*

Lesson ______ Time [:] Money ______ Secret Number ______

Lesson ______ Time [:] Money ______ Secret Number ______

Lesson ______ Time [:] Money ______ Secret Number ______

Lesson ______ Time [:] Money ______ Secret Number ______

Name ______________________________

Meeting Worksheet

***Saxon** Math 2 (first used in **Lesson 18**)*

Lesson ______ Time [:] Money ______ Secret Number ______

Lesson ______ Time [:] Money ______ Secret Number ______

Lesson ______ Time [:] Money ______ Secret Number ______

Lesson ______ Time [:] Money ______ Secret Number ______

Name ______________________

Meeting Worksheet

Saxon *Math 2 (first used in* ***Lesson 18****)*

Lesson ______ Time [:] Money ______ Secret Number ______

Lesson ______ Time [:] Money ______ Secret Number ______

Lesson ______ Time [:] Money ______ Secret Number ______

Lesson ______ Time [:] Money ______ Secret Number ______

Name ______________________

Meeting Worksheet

***Saxon** Math 2 (first used in **Lesson 18**)*

Lesson ______ Time [:] Money ______ Secret Number ______

Lesson ______ Time [:] Money ______ Secret Number ______

Lesson ______ Time [:] Money ______ Secret Number ______

Lesson ______ Time [:] Money ______ Secret Number ______

Name ______________________

Meeting Worksheet

Saxon *Math 2 (first used in* ***Lesson 18****)*

Lesson ______ Time [:] Money ______ Secret Number ______

Lesson ______ Time [:] Money ______ Secret Number ______

Lesson ______ Time [:] Money ______ Secret Number ______

Lesson ______ Time [:] Money ______ Secret Number ______

Name ______________________

Meeting Worksheet

***Saxon** Math 2 (first used in **Lesson 18**)*

Lesson ______ Time [:] Money ______ Secret Number ______

Lesson ______ Time [:] Money ______ Secret Number ______

Lesson ______ Time [:] Money ______ Secret Number ______

Lesson ______ Time [:] Money ______ Secret Number ______

Name ____________________

Meeting Worksheet

***Saxon** Math 2 (first used in **Lesson 18**)*

Lesson ______ Time [:] Money ______ Secret Number ______

Lesson ______ Time [:] Money ______ Secret Number ______

Lesson ______ Time [:] Money ______ Secret Number ______

Lesson ______ Time [:] Money ______ Secret Number ______

Name ________________________________

Meeting Worksheet

Saxon *Math 2 (first used in* ***Lesson 18****)*

Lesson ______ Time [:] Money ______ Secret Number ______

Lesson ______ Time [:] Money ______ Secret Number ______

Lesson ______ Time [:] Money ______ Secret Number ______

Lesson ______ Time [:] Money ______ Secret Number ______

Name ______________________________

Meeting Worksheet

***Saxon** Math 2 (first used in **Lesson 18**)*

Lesson ________ Time [:] Money ________ Secret Number ________

Lesson ________ Time [:] Money ________ Secret Number ________

Lesson ________ Time [:] Money ________ Secret Number ________

Lesson ________ Time [:] Money ________ Secret Number ________

Name ______________________________

Meeting Worksheet

***Saxon** Math 2 (first used in **Lesson 18**)*

Lesson ________ Time [:] Money ________ Secret Number ________

Lesson ________ Time [:] Money ________ Secret Number ________

Lesson ________ Time [:] Money ________ Secret Number ________

Lesson ________ Time [:] Money ________ Secret Number ________

Meeting Worksheet

Saxon Math 2 (first used in **Lesson 18**)

Name ______________________

Lesson ______ Time [:] Money ______ Secret Number ______

Lesson ______ Time [:] Money ______ Secret Number ______

Lesson ______ Time [:] Money ______ Secret Number ______

Lesson ______ Time [:] Money ______ Secret Number ______

Name ______________________________

Meeting Worksheet

Saxon *Math 2 (first used in* ***Lesson 18****)*

Lesson ________ Time [:] Money ________ Secret Number ________

Lesson ________ Time [:] Money ________ Secret Number ________

Lesson ________ Time [:] Money ________ Secret Number ________

Lesson ________ Time [:] Money ________ Secret Number ________

Name ______________________

Meeting Worksheet

***Saxon** Math 2 (first used in **Lesson 18**)*

Lesson ______ Time [:] Money ______ Secret Number ______

Lesson ______ Time [:] Money ______ Secret Number ______

Lesson ______ Time [:] Money ______ Secret Number ______

Lesson ______ Time [:] Money ______ Secret Number ______

Name ___________________________

Meeting Worksheet

***Saxon** Math 2 (first used in **Lesson 18**)*

Lesson ______ Time [:] Money ______ Secret Number ______

Lesson ______ Time [:] Money ______ Secret Number ______

Lesson ______ Time [:] Money ______ Secret Number ______

Lesson ______ Time [:] Money ______ Secret Number ______

Name ____________________

Meeting Worksheet

***Saxon** Math 2 (first used in **Lesson 18**)*

Lesson ______ Time [:] Money ______ Secret Number ______

Lesson ______ Time [:] Money ______ Secret Number ______

Lesson ______ Time [:] Money ______ Secret Number ______

Lesson ______ Time [:] Money ______ Secret Number ______

Name ______________________________

Date ______________________________

Guided Class Practice 2A

***Saxon** Math 2 (for use with **Lesson 2**)*

1. What day of the week is it today?

____

2. Write the letter **e** to the right of the **n.**
Write the letter **o** to the left of the **n.**

_____ n _____ _____

3. Use a red crayon to color these numbers on the chart.
Cross off each number after you color it.

14, 1, 28, 10, 17, 5, 34, 12, 23,
50, 19, 46, 37, 6, 45, 32, 39, 41

1	2	3	4	5	6	7	8	9	10
11	12	13	14	15	16	17	18	19	20
21	22	23	24	25	26	27	28	29	30
31	32	33	34	35	36	37	38	39	40
41	42	43	44	45	46	47	48	49	50

4. What number is one more than **28?** ________

What number is one less than **37?** ________

Name ____________________________

Date ____________________________

Homework 2B

***Saxon** Math 2 (for use with **Lesson 2**)*

1. Read these numbers to someone.

39, 18, 12, 22, 40, 48

2. Write the letter **t** to the left of the **w**.
Write the letter **o** to the right of the **w**.

_____ W _____ _____

3. Use a red crayon to color these numbers on the chart.
Cross off each number after you color it.

22, 14, 37, 8, 23, 34, 42, 17, 48, 6,
12, 44, 27, 46, 4, 32, 47, 7, 24, 2

1	2	3	4	5	6	7	8	9	10
11	12	13	14	15	16	17	18	19	20
21	22	23	24	25	26	27	28	29	30
31	32	33	34	35	36	37	38	39	40
41	42	43	44	45	46	47	48	49	50

Dear Parent/Guardian,

During the coming year your child will participate in a wide variety of mathematics activities using the *Saxon Math 2* program. Your child will learn through hands-on experiences, discussions, explorations, and oral/written practice.

While each day's activities will be varied, each lesson will have a standard four-part format:

1. **The Meeting** is a time when we practice everyday skills. The children solve a problem of the day and learn about the calendar, a daily number or shape pattern, temperature, attendance graph, time, money, and fact families.
2. **Fact Practice** helps children master number facts by practicing fact strategies using fact cards, games and activities, Learning Wrap-Ups®, and fact sheets. Your child will also practice the number facts at home using fact sheets.
3. **New Concepts** are presented in each lesson through discussion and hands-on experiences that allow your child to be actively involved in learning.
4. **Written Practice** reinforces new concepts from the lesson, as well as from previous lessons. The children are guided in class as they complete and correct Side A of a practice sheet. Your child will complete Side B as homework.

   Please assist your child by reading the problems on Side B, if necessary. Allow your child to arrive at the answers independently. Check your child's work and help your child correct mistakes. If you help your child with a problem, please circle the problem number to let me know that this is a difficult question. **It is important that your child return the homework the next day.**

Assessments, both written and oral, include skills your child has been practicing throughout the year and will help me determine what additional review is necessary. I will share with you my observations about your child's progress.

I look forward to working with you and your child this year. Please contact me if you have any questions about the program or about your child's progress.

Sincerely,

Name ______________________________

Date ______________________________

Guided Class Practice **3A**

***Saxon** Math 2 (for use with **Lesson 3**)*

1. Use the class birthday graph to answer these questions.

How many children have birthdays in August? ________

How many children have birthdays in September? ________

2. What number is one less than **13?** ________

What number is one more than **20?** ________

3. Count from one to twenty. Write the numbers.

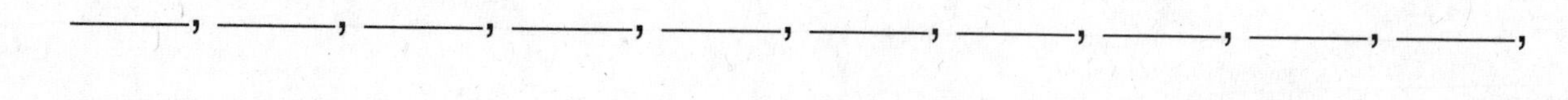

4. Write the digital time.

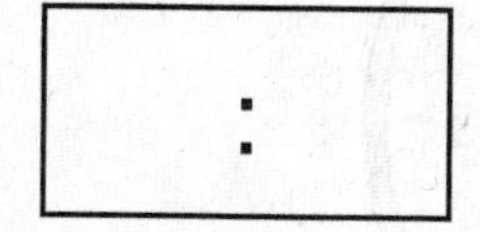

5. Write the letter **s** to the left of the **c.**
Write the letter **h** to the right of the **c.**
Write the letter **o** to the left of the **l.**

______ __c__ ______ __o__ ______ __l__

Name ______________________________

Date ______________________________

1. What will be tomorrow's date?

____

2. What number is one less than **25?** ________

What number is one more than **16?** ________

3. Count backward from twenty to one. Write the numbers.

20, ___, ___, ___, ___, ___, ___, ___, ___, ___,
___, ___, ___, ___, ___, ___, ___, ___, ___, ___

4. Write the digital time.

[:   ]

5. Write the letter **u** to the left of the **s.**
Write the letter **e** to the right of the **s.**
Write the letter **h** to the left of the **o.**

___ o ___ s ___

Name ______________________________

Date ______________________________

Guided Class Practice 4A

*Saxon Math 2 (for use with **Lesson 4**)*

1. Fill in the missing days of the week.

Sunday, ______________, Tuesday, ______________,

Thursday, ______________, Saturday

2. Which letter is on the right? ________

Which letter is in the middle? ________

Which letter is on the left? ________

e i p

3. Use the class birthday graph to answer these questions.

How many birthdays are in June? ________

Which month has the most birthdays? ______________

4. Count by 10's. Write the numbers.

10, 20, ____, ____, ____, ____, ____, ____, ____, ____

5. Fill in the missing numbers.

1	2	3	4	5	6	7	8	9	10
21	22	23	24	25	26	27	28	29	30
41	42	43	44	45	46	47	48	49	50

Name ____________________________

Date ____________________________

Saxon Math 2 (for use with **Lesson 4**)

1. Fill in the missing days of the week.

__________, Monday, __________, Wednesday,

__________, Friday, __________

2. Which letter is on the left? ______

Which letter is on the right? ______

Which letter is in the middle? ______

b y o

3. Fill in the missing numbers.

1, 2, ____, ____, ____, ____, ____, 8, 9, ____,

____, ____, 13, ____, ____, 16, ____, ____, 19, ____

4. Count backward by 10's. Write the numbers.

100, 90, 80, ____, ____, ____, ____, ____, ____, ____

5. Fill in the missing numbers.

1	2	3	4		6	7	8	9	
11	12	13	14		16	17	18	19	
21	22	23	24		26	27	28	29	
31	32	33	34		36	37	38	39	
41	42	43	44		46	47	48	49	

Name ___________________________

Lesson Worksheet 5A

Saxon Math 2 (for use with **Lesson 5**)

A.

$0 + 0$	$1 + 1$	$2 + 2$	$3 + 3$	$4 + 4$
$5 + 5$	$6 + 6$	$7 + 7$	$8 + 8$	$9 + 9$

B.

$6 + 6$	$2 + 2$	$1 + 1$	$9 + 9$	$4 + 4$
$0 + 0$	$8 + 8$	$5 + 5$	$3 + 3$	$7 + 7$

C.

$2 + \square = 4$	$8 + \square = 16$	$3 + \square = 6$	$5 + \square = 10$	$9 + \square = 18$

Dear Parent,

Today we practiced the doubles addition facts. Your child will be tested next week on the new addition facts listed below.

$$\begin{array}{r} 0 \\ +\ 0 \\ \hline \end{array} \quad \begin{array}{r} 1 \\ +\ 1 \\ \hline \end{array} \quad \begin{array}{r} 2 \\ +\ 2 \\ \hline \end{array} \quad \begin{array}{r} 3 \\ +\ 3 \\ \hline \end{array} \quad \begin{array}{r} 4 \\ +\ 4 \\ \hline \end{array}$$

$$\begin{array}{r} 5 \\ +\ 5 \\ \hline \end{array} \quad \begin{array}{r} 6 \\ +\ 6 \\ \hline \end{array} \quad \begin{array}{r} 7 \\ +\ 7 \\ \hline \end{array} \quad \begin{array}{r} 8 \\ +\ 8 \\ \hline \end{array} \quad \begin{array}{r} 9 \\ +\ 9 \\ \hline \end{array}$$

During the next four days, your child will practice these facts in class both orally and in writing. The back of each day's fact sheet will contain the fact practice homework.

Tonight's fact practice is the following:

1. Ask your child to point to the problems in order and to say the answers.
2. Point to the problems in random order and ask your child to say the answers.

Keep this paper at home to use for practice.

Name ______________________

Date ______________________

Guided Class Practice **5A**

***Saxon** Math 2 (for use with **Lesson 5**)*

1. What day of the week is it today? ______________

   What day of the week was it yesterday? ______________

2. Write the letter **h** to the left of the **o.**
   Write the letter **t** to the right of the **o.**

   ____ o ____

3. Fill in the missing numbers.

   1, 2, ____, 4, ____, ____, 7, ____, 9, ____,
   ____, ____, 13, ____, ____, ____, ____, 18, ____, ____

4. What number is one less than **26?** ______

   What number is one more than **43?** ______

5. Use a red crayon to color these numbers on the chart.
   Cross off each number after you color it.

   23, 42, 37, 8, 29, 4, 2, 17, 10, 27, 5,
   41, 43, 45, 3, 28, 9, 1, 33, 47, 13, 44, 7

1	2	3	4	5	6	7	8	9	10
11	12	13	14	15	16	17	18	19	20
21	22	23	24	25	26	27	28	29	30
31	32	33	34	35	36	37	38	39	40
41	42	43	44	45	46	47	48	49	50

Name ______________________________

Date ______________________________

Homework 5B

Saxon *Math 2 (for use with* ***Lesson 5****)*

1. What will be tomorrow's date?

____

2. Write the letter **t** to the left of the **o**.
Write the letter **e** to the right of the **o**.

____ o ____

3. Fill in the missing numbers.

1, ____, 3, ____, ____, ____, ____, 8, ____, ____,

____, 12, ____, ____, ____, 16, ____, ____, ____, ____

4. What number is one more than **76?** ________

What number is one less than **53?** ________

5. Fill in the missing numbers.

1	2		4	5		7		9	10
		13		15			18		
21		23			26			29	
	32			35				39	
41			44		46	47			50

Name ______________________________

Set 1: Doubles

2 + 2	6 + 6	1 + 1	4 + 4	9 + 9
7 + 7	0 + 0	8 + 8	5 + 5	3 + 3
1 + 1	9 + 9	5 + 5	2 + 2	8 + 8
6 + 6	3 + 3	0 + 0	7 + 7	4 + 4
9 + 9	5 + 5	8 + 8	6 + 6	7 + 7

Name ______________________________

Fact Homework 6B

Saxon *Math 2 (for use with* ***Lesson 6****)*

Set 1: Doubles

1. Read the answers to someone.
2. Write the answers.
3. Ask someone to correct your paper. Corrected by ____________________

1 + 1	9 + 9	4 + 4	2 + 2	7 + 7
8 + 8	5 + 5	0 + 0	3 + 3	6 + 6
7 + 7	1 + 1	9 + 9	0 + 0	4 + 4
2 + 2	6 + 6	5 + 5	8 + 8	3 + 3
9 + 9	4 + 4	7 + 7	2 + 2	5 + 5

Name ________________________

Date ________________________

Guided Class Practice 6A

***Saxon** Math 2 (for use with **Lesson 6**)*

1. Fill in the missing days of the week.

Tuesday, ____________, ____________, Friday

2. Draw a square to the left of the circle.
Draw a triangle to the right of the circle.

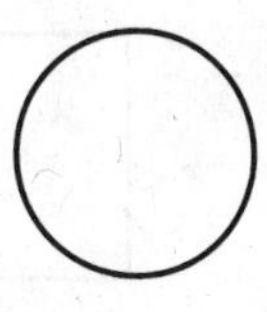

3. Use the class birthday graph to answer these questions.

How many birthdays are in February? ______

Which months have exactly two birthdays?

____

____

4. Count backward by 10's. Fill in the missing numbers.

100, 90, 80, ____, ____, ____, ____, ____, ____, ____

5. Write the digital time.

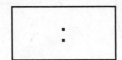

Name ______________________

Date ______________________

Homework 6B

***Saxon** Math 2 (for use with **Lesson 6**)*

1. Fill in the missing days of the week.

__________, Wednesday, Thursday, __________

2. Draw a circle to the right of the square.
Draw a triangle to the left of the square.

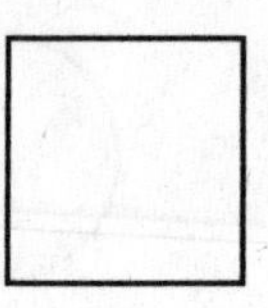

3. Fill in the missing numbers.

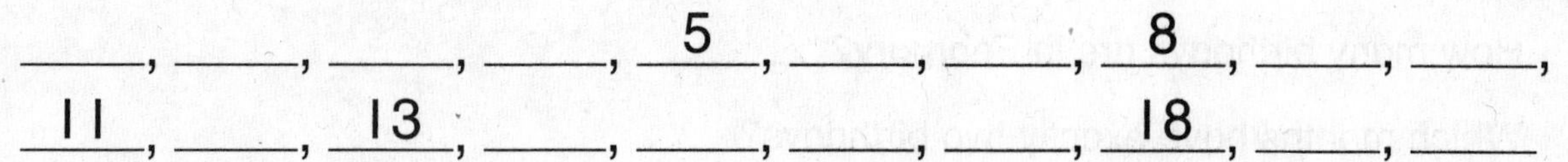

___, ___, ___, ___, 5, ___, ___, 8, ___, ___,

11, ___, 13, ___, ___, ___, ___, 18, ___, ___

4. Count by 10's. Fill in the missing numbers.

10, 20, 30, ___, ___, ___, ___, ___, ___, ___

5. Write the digital time.

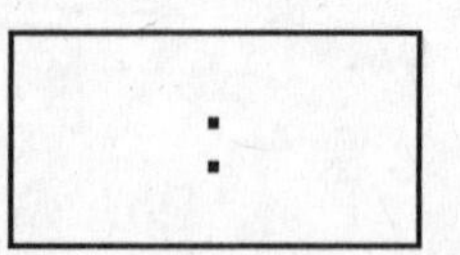

Name ____________________

Set 1: Doubles

6 + 6	2 + 2	9 + 9	0 + 0	4 + 4
8 + 8	1 + 1	7 + 7	5 + 5	3 + 3
7 + 7	0 + 0	8 + 8	1 + 1	6 + 6
4 + 4	9 + 9	2 + 2	5 + 5	3 + 3
7 + 7	1 + 1	6 + 6	3 + 3	8 + 8

Name ______________________

Fact Homework 7B

***Saxon** Math 2 (for use with **Lesson 7**)*

Set 1: Doubles

1. Read the answers to someone.
2. Write the answers.
3. Ask someone to correct your paper. Corrected by ______________________

$4 + 4$	$7 + 7$	$3 + 3$	$9 + 9$	$2 + 2$
$5 + 5$	$8 + 8$	$1 + 1$	$6 + 6$	$0 + 0$
$3 + 3$	$6 + 6$	$9 + 9$	$4 + 4$	$1 + 1$
$7 + 7$	$0 + 0$	$8 + 8$	$5 + 5$	$2 + 2$
$4 + 4$	$9 + 9$	$6 + 6$	$3 + 3$	$8 + 8$

Name ______________________________

Date ______________________________

Guided Class Practice 7A

***Saxon** Math 2 (for use with **Lesson 7**)*

1. Fill in the missing days of the week.

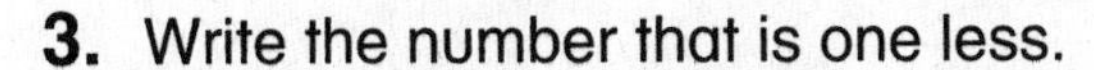

______________, Monday, ______________

2. Write an **i** on the second line.
Write a **t** on the fifth line.
Write a **g** on the third line.
Write an **h** on the fourth line.
Write an **r** on the first line.

_____ _____ _____ _____ _____

3. Write the number that is one less.

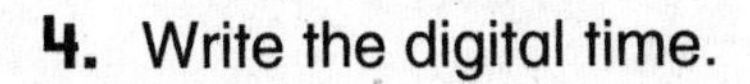

_____, 28 _____, 16

4. Write the digital time.

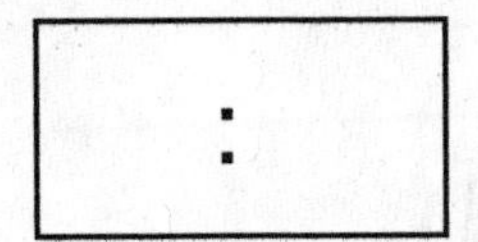

5. Find the sums.

4	7	3	9	6	2	8
+ 4	+ 7	+ 3	+ 9	+ 6	+ 2	+ 8

Name ______________________________

Date ______________________________

Homework 7B

***Saxon** Math 2 (for use with **Lesson 7**)*

1. Fill in the missing days of the week.

Sunday, ______________, ______________

2. Write an **e** on the third line.
Write an **a** on the fourth line.
Write a **g** on the first line.
Write a **t** on the fifth line.
Write an **r** on the second line.

_____ _____ _____ _____ _____

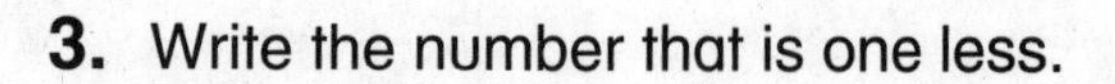

3. Write the number that is one less.

_____, 49 _____, 12

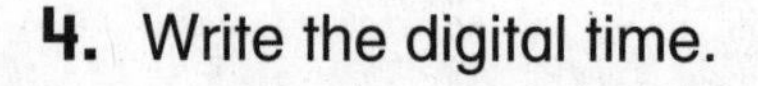

4. Write the digital time.

___ : ___

5. Find the sums.

5	1	8	3	6	10	7
+ 5	+ 1	+ 8	+ 3	+ 6	+ 10	+ 7

Name ______________________

Set 1: Doubles

A.

$4 + 4 =$ ____	$5 + 5 =$ ____	$1 + 1 =$ ____
$9 + 9 =$ ____	$0 + 0 =$ ____	$7 + 7 =$ ____
$2 + 2 =$ ____	$8 + 8 =$ ____	$3 + 3 =$ ____
$6 + 6 =$ ____	$1 + 1 =$ ____	$9 + 9 =$ ____
$4 + 4 =$ ____	$7 + 7 =$ ____	$5 + 5 =$ ____
$3 + 3 =$ ____	$2 + 2 =$ ____	$8 + 8 =$ ____
$0 + 0 =$ ____	$9 + 9 =$ ____	$4 + 4 =$ ____

B.

$2 + \square = 4$	$5 + \square = 10$	$9 + \square = 18$
$\square + 1 = 2$	$\square + 4 = 8$	$\square + 0 = 0$

Name ______________________

Fact Homework 8B

***Saxon** Math 2 (for use with **Lesson 8**)*

Set 1: Doubles

A. 1. Read the answers to someone.
2. Write the answers.
3. Ask someone to correct your paper. Corrected by ______________

3 + 3 = ____ 7 + 7 = ____ 0 + 0 = ____

4 + 4 = ____ 9 + 9 = ____ 2 + 2 = ____

8 + 8 = ____ 1 + 1 = ____ 5 + 5 = ____

6 + 6 = ____ 0 + 0 = ____ 3 + 3 = ____

9 + 9 = ____ 7 + 7 = ____ 4 + 4 = ____

2 + 2 = ____ 8 + 8 = ____ 1 + 1 = ____

6 + 6 = ____ 5 + 5 = ____ 0 + 0 = ____

B. Fill in the missing numbers.

3 + ☐ = 6 7 + ☐ = 14 2 + ☐ = 4

☐ + 8 = 16 ☐ + 1 = 2 ☐ + 6 = 12

Name ______________________________

Date ______________________________

Guided Class Practice 8A

***Saxon** Math 2 (for use with **Lesson 8**)*

1. Use the class birthday graph to answer these questions.

What is the first month of the year? ____________________

How many children have a birthday in that month? ________

2. One of these is my dog.
Use the clues to find my dog.
He is not third. Cross out that dog.
He is not on the left. Cross out that dog.
Circle my dog.

3. Continue the repeating pattern.
Color the circles red.
Color the squares green.

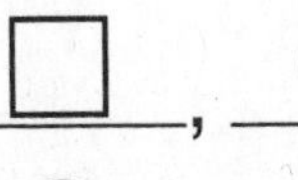 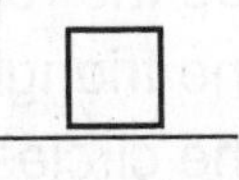 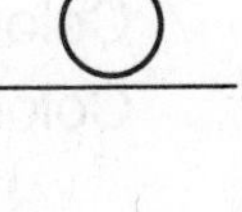

_____, _____, _____, _____, _____,
_____, _____, _____, _____, _____,
_____, _____, _____, _____, _____

4. Write the digital time.

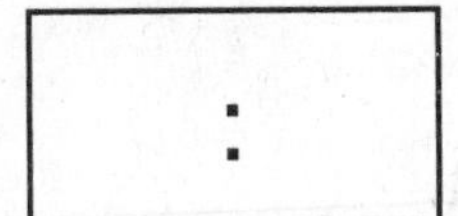

5. Circle the greater number. 24 17

6. Fill in the missing numbers.

1		3		5		7		9	
11		13		15		17		19	
21		23		25		27		29	

Name ______________________

Date ______________________

Homework 8B

Saxon Math 2 (for use with Lesson 8)

1. What will be tomorrow's date?

____

2. One of these is my grandmother's dog.
Use the clues to find her dog.
She is not the second. Cross out that dog.
She is not on the left. Cross out that dog.
Circle her dog.

3. Continue the repeating pattern.
Color the triangles blue.
Color the circles yellow.

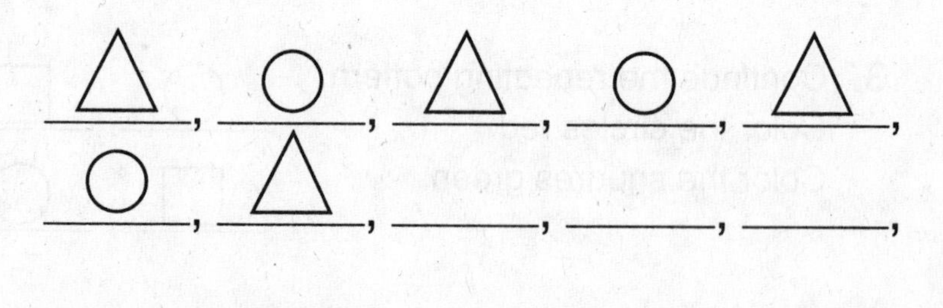

4. Write the digital time.

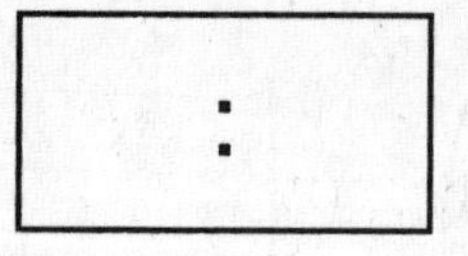

5. Circle the greater number.

38 43

6. Fill in the missing numbers.

	2		4		6		8		10
	12		14		16		18		20
	22		24		26		28		30

This page may not be reproduced without permission of Harcourt Achieve Inc.

Name ______________________________

Class Fact Practice 9A

***Saxon** Math 2 (for use with **Lesson 9**)*

Set 1: Doubles

3 + 3	9 + 9	5 + 5	7 + 7	1 + 1
4 + 4	6 + 6	0 + 0	8 + 8	2 + 2
1 + 1	6 + 6	9 + 9	5 + 5	3 + 3
0 + 0	8 + 8	2 + 2	7 + 7	4 + 4
5 + 5	2 + 2	6 + 6	3 + 3	9 + 9

Name ______________________________

Fact Homework 9B

***Saxon** Math 2 (for use with **Lesson 9**)*

Set 1: Doubles

1. Read the answers to someone.
2. Write the answers.
3. Ask someone to correct your paper. Corrected by ______________

$5 + 5$	$9 + 9$	$1 + 1$	$6 + 6$	$3 + 3$
$0 + 0$	$7 + 7$	$4 + 4$	$2 + 2$	$8 + 8$
$1 + 1$	$6 + 6$	$3 + 3$	$7 + 7$	$5 + 5$
$9 + 9$	$0 + 0$	$8 + 8$	$2 + 2$	$4 + 4$
$7 + 7$	$1 + 1$	$6 + 6$	$4 + 4$	$8 + 8$

Name ______________________________

Date ______________________________

1. Use the class birthday graph to answer the questions.

How many children have birthdays in August? ________

What is the fourth month of the year? ______________________

How many children have birthdays in that month? ________

2. Count by 10's. Fill in the missing numbers.

10, _____, _____, _____, _____, _____, _____, _____, _____, _____

3. The teddy bears are in line next to the toy chest.

Color the first teddy bear blue.
Color the fifth teddy bear green.
Color the second teddy bear yellow.
Color the third teddy bear red.

In what position is the bear that is not colored? ______________________

4. Circle the number that is greater. 43 49

5. Circle each shape that has equal-size pieces.

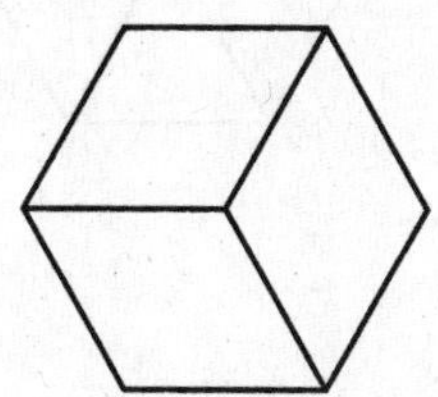

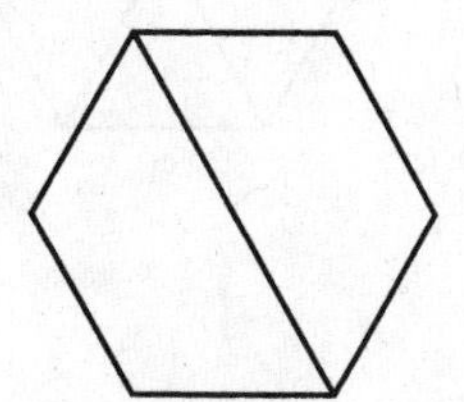

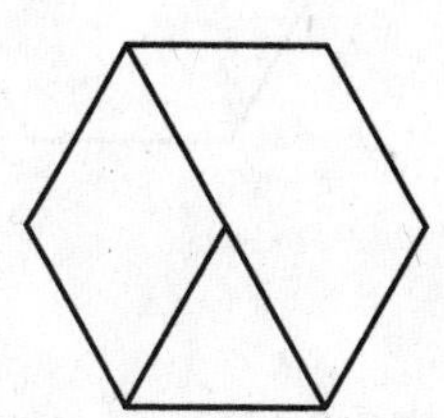

6. Find the sums.

7 + 7 = ____ 4 + 4 = ____ 8 + 8 = ____ 3 + 3 = ____

Name ______________________________

Date ______________________________

1. What will be tomorrow's date?

____

2. Count backward by 10's. Fill in the missing numbers.

100, ____, ____, ____, ____, ____, ____, ____, ____, ____

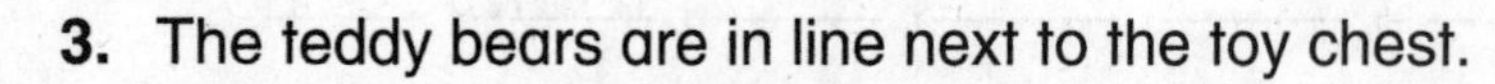

3. The teddy bears are in line next to the toy chest.

Color the first teddy bear green.
Color the fourth teddy bear yellow.
Color the third teddy bear red.
Color the fifth teddy bear blue.

In what position is the bear that is not colored? ____________

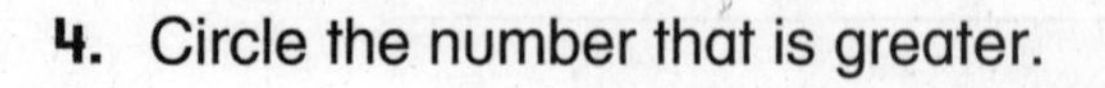

4. Circle the number that is greater.

58 54

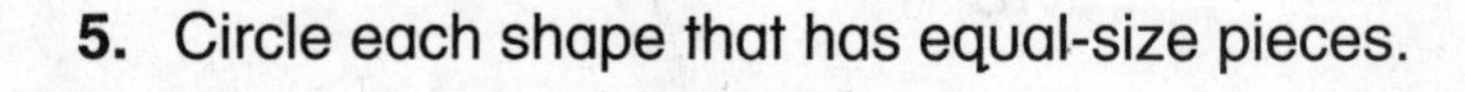

5. Circle each shape that has equal-size pieces.

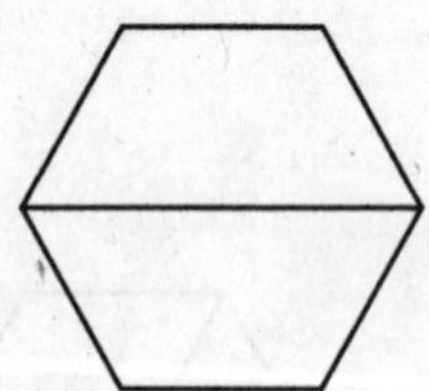

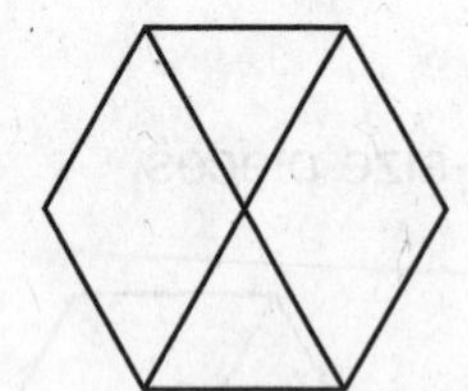

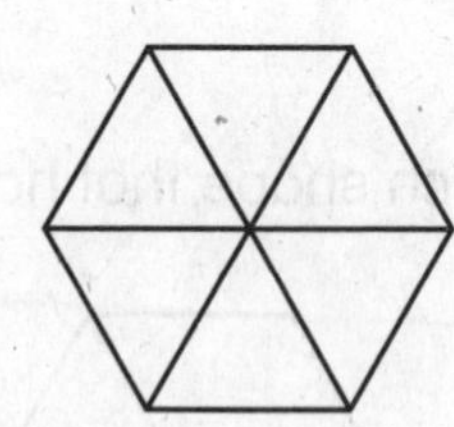

6. Find the sums.

$5 + 5 =$ ____ $9 + 9 =$ ____ $2 + 2 =$ ____ $6 + 6 =$ ____

Name ________________ Score ________

Fact Assessment 1

***Saxon** Math 2 (for use with **Lesson 10-1**)*

Set 1: Doubles

2 + 2	5 + 5	9 + 9	1 + 1	6 + 6
3 + 3	7 + 7	0 + 0	4 + 4	8 + 8
5 + 5	1 + 1	3 + 3	9 + 9	2 + 2
6 + 6	0 + 0	8 + 8	4 + 4	7 + 7
9 + 9	2 + 2	1 + 1	5 + 5	4 + 4

Name ______________________

Lesson Worksheet 10-1A

***Saxon** Math 2 (for use with **Lesson 10-1**)*

A.

1 + 0 = 1; 0 + 1 = 1	2 + 0; +	3 + 0; +
4 + 0; +	5 + 0; +	6 + 0; +
7 + 0; +	8 + 0; +	9 + 0; +

B.

2 + 1 = 3; 1 + 2 = 3	3 + 1; +	4 + 1; +
5 + 1; +	6 + 1; +	7 + 1; +
8 + 1; +	9 + 1; +	

Dear Parent,

Today we practiced the adding 0 and adding 1 facts. Your child will be tested next week on the new addition facts listed below.

1 + 0	2 + 0	3 + 0	4 + 0	5 + 0	6 + 0	7 + 0
8 + 0	9 + 0	0 + 1	0 + 2	0 + 3	0 + 4	0 + 5
0 + 6	0 + 7	0 + 8	0 + 9	2 + 1	3 + 1	4 + 1
5 + 1	6 + 1	7 + 1	8 + 1	9 + 1	1 + 2	1 + 3
1 + 4	1 + 5	1 + 6	1 + 7	1 + 8	1 + 9	

During the next four days, your child will practice these facts in class both orally and in writing. The back of each day's fact sheet will contain the fact practice homework.

Tonight's fact practice is the following:

1. Ask your child to point to the problems in order and to say the answers.
2. Point to the problems in random order and ask your child to say the answers.

Keep this paper at home to use for practice.

Name ______________________________

Date ______________________________

Problem-Solving Worksheet 10A

***Saxon** Math 2 (for use with **Lesson 10-1**)*

Understand	Plan	Solve	Check

Use Logical Reasoning

Act It Out

Maria, Jose, and Zenia are standing in line at the door. Jose is last. Maria is not first. Show who is second in line.

Who is second in line? ______________________

Name ______________________

Performance Task Worksheet 10B

***Saxon** Math 2 (for use with **Lesson 10-1**)*

Understand	Plan	Solve	Check

Carson put her books in a stack on her desk. The red book is not on top. The green book is in the middle. The yellow book is above the green book. Show which book is at the bottom of the stack.

What color book is at the bottom of the stack? ______________________

Circle the problem-solving strategies you used to solve this problem.

Act It Out ***Use Logical Reasoning***

Explain how you got your answer: ______________________

Name ______________________________

Date ______________________________

1. Use the class birthday graph to answer the questions.

   How many children have birthdays in September? ________

   Which month has the most birthdays? ________________

2. Continue the repeating pattern.

   □, ○, □, ○, ______, ______, ______, ______, ______, ______

3. What number is one more than **34**? ________

   What number is one less than **54**? ________

4. Write the digital time.

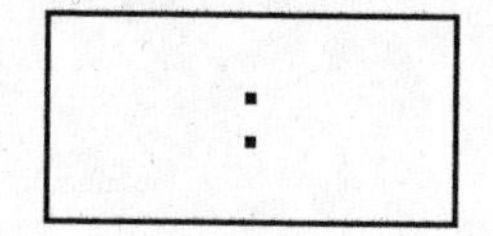

5. Fill in the missing numbers.

1	2			5			8		
		13				17			20
21			24		26				

6. Find the sums.

$2 + 2 =$ ______ $5 + 5 =$ ______ $8 + 8 =$ ______ $3 + 3 =$ ______

$7 + 7 =$ ______ $4 + 4 =$ ______ $9 + 9 =$ ______ $6 + 6 =$ ______

Name ______________________

How many blocks of each color did you use?

yellow ________

red ________

blue ________

green ________

orange ________

tan ________

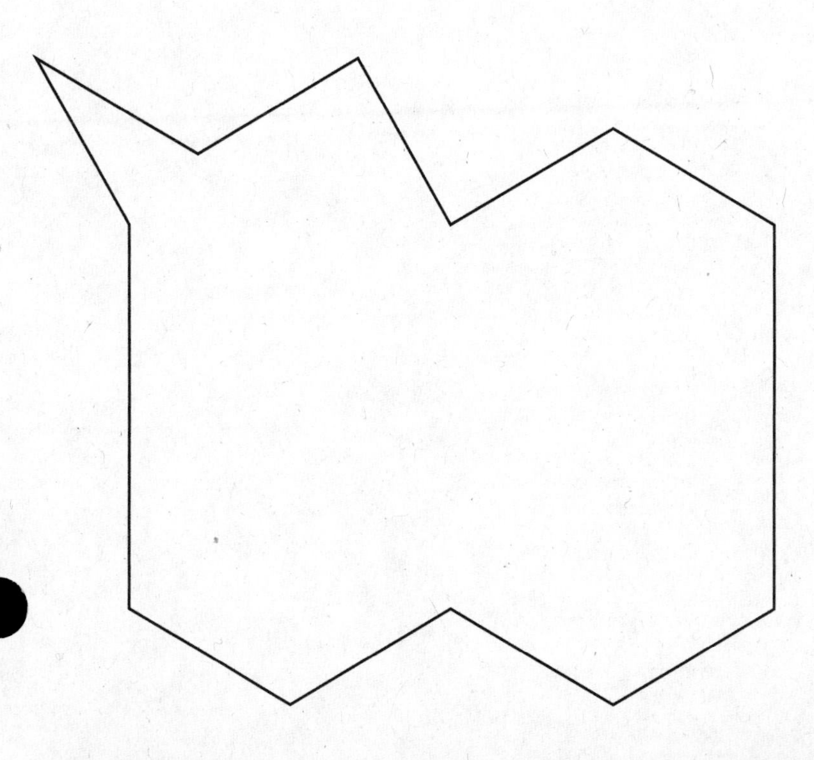

How many blocks of each color did you use?

yellow ________

red ________

blue ________

green ________

orange ________

tan ________

Name ____________________________

Set 2: Adding 1 and 0; Doubles

9 + 0	4 + 1	1 + 7	6 + 6	0 + 3
8 + 1	7 + 7	9 + 0	1 + 2	0 + 7
1 + 6	0 + 4	8 + 8	5 + 1	2 + 0
1 + 3	5 + 0	9 + 1	0 + 8	9 + 9
0 + 2	1 + 4	6 + 0	5 + 5	1 + 0

Name ______________________________

Set 2: Adding 1 and 0; Doubles

1. Read the answers to someone.
2. Write the answers.
3. Ask someone to correct your paper. Corrected by ____________________

$$\begin{array}{r}3\\+\ 0\\\hline\end{array}\quad\begin{array}{r}9\\+\ 1\\\hline\end{array}\quad\begin{array}{r}0\\+\ 5\\\hline\end{array}\quad\begin{array}{r}6\\+\ 6\\\hline\end{array}\quad\begin{array}{r}1\\+\ 8\\\hline\end{array}$$

$$\begin{array}{r}1\\+\ 4\\\hline\end{array}\quad\begin{array}{r}8\\+\ 0\\\hline\end{array}\quad\begin{array}{r}3\\+\ 3\\\hline\end{array}\quad\begin{array}{r}1\\+\ 7\\\hline\end{array}\quad\begin{array}{r}0\\+\ 1\\\hline\end{array}$$

$$\begin{array}{r}7\\+\ 7\\\hline\end{array}\quad\begin{array}{r}2\\+\ 1\\\hline\end{array}\quad\begin{array}{r}4\\+\ 0\\\hline\end{array}\quad\begin{array}{r}1\\+\ 3\\\hline\end{array}\quad\begin{array}{r}0\\+\ 6\\\hline\end{array}$$

$$\begin{array}{r}7\\+\ 1\\\hline\end{array}\quad\begin{array}{r}9\\+\ 0\\\hline\end{array}\quad\begin{array}{r}1\\+\ 5\\\hline\end{array}\quad\begin{array}{r}8\\+\ 8\\\hline\end{array}\quad\begin{array}{r}3\\+\ 1\\\hline\end{array}$$

$$\begin{array}{r}0\\+\ 8\\\hline\end{array}\quad\begin{array}{r}6\\+\ 1\\\hline\end{array}\quad\begin{array}{r}9\\+\ 9\\\hline\end{array}\quad\begin{array}{r}0\\+\ 2\\\hline\end{array}\quad\begin{array}{r}1\\+\ 9\\\hline\end{array}$$

Name ______________________________

Date ______________________________

Guided Class Practice 11A

***Saxon** Math 2 (for use with **Lesson 11**)*

1. One of these is my favorite color. Use the clues to find my favorite color.

Red | Blue | Yellow | Green

It is not third. Cross out that color.
It is not on the right. Cross out that color.
It is not first. Cross out that color.
Circle my favorite color.

2. Circle the smaller number. 28 34

3. Continue the repeating pattern.
Color the triangles green.
Color the squares yellow.
Color the circles red.

△, □, ○, △, □, ○, ____, ____, ____, ____, ____, ____

4. What number is one more than **63?** ________

What number is one less than **47?** ________

5. Fill in the missing numbers.

Which number did you write in the square with the **A?** ________

Which number did you write in the square with the **B?** ________

1	2			5					10
		13							
			A	26					
31			34						40
41				B				49	50

Name ________________________

Date ________________________

Homework 11B

***Saxon** Math 2 (for use with **Lesson 11**)*

1. One of these is my sister's favorite color. Use the clues to find her favorite color.

Red | Blue | Yellow | Green

It is not third. Cross out that color.
It is not on the right. Cross out that color.
It is not second. Cross out that color.
Circle my sister's favorite color.

2. Circle the smaller number. 29 23

3. Continue the repeating pattern.
Color the circles blue.
Color the squares yellow.

○, □, ○, □, ○, □, ____, ____, ____, ____, ____, ____

4. What number is one more than **25?** ______

What number is one less than **73?** ______

5. Fill in the missing numbers.

Which number did you write in the square with the **A?** ______

Which number did you write in the square with the **B?** ______

1	2			5					10
11							B		
						27			
31									
41		A			46				50

Name ______________________________

Class Fact Practice 12A

Saxon Math 2 (for use with Lesson 12)

Set 2: Adding 1 and 0; Doubles

$4 + 1$	$0 + 8$	$9 + 9$	$1 + 0$	$6 + 1$
$2 + 1$	$3 + 3$	$0 + 6$	$1 + 5$	$7 + 7$
$1 + 9$	$3 + 0$	$1 + 6$	$0 + 0$	$1 + 7$
$9 + 0$	$1 + 3$	$0 + 2$	$8 + 1$	$5 + 0$
$1 + 1$	$5 + 1$	$7 + 0$	$8 + 8$	$0 + 4$

Name ____________________

Set 2: Adding 1 and 0; Doubles

1. Read the answers to someone.
2. Write the answers.
3. Ask someone to correct your paper. Corrected by ____________________

$3 + 1$	$5 + 5$	$1 + 6$	$2 + 0$	$7 + 7$
$0 + 5$	$1 + 8$	$2 + 2$	$9 + 1$	$6 + 0$
$1 + 2$	$0 + 7$	$6 + 6$	$0 + 3$	$5 + 1$
$9 + 9$	$4 + 0$	$7 + 1$	$4 + 4$	$0 + 1$
$1 + 4$	$3 + 3$	$8 + 0$	$0 + 0$	$1 + 9$

Name ____________________

1 hour ago

:

now

:

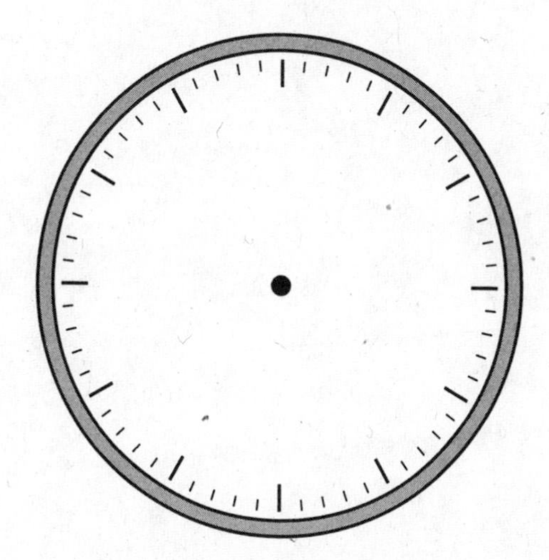

1 hour from now

:

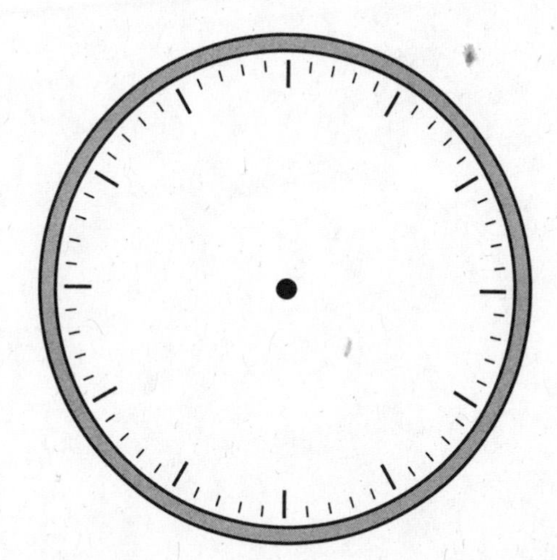

Name ______________________________

Date ______________________________

Guided Class Practice 12A

***Saxon** Math 2 (for use with **Lesson 12**)*

1. There were 10 children at the bus stop. Two more children joined them. How many children are at the bus stop now? What type of story problem is this?

Circle one: Some, some more Some, some went away

2. Circle the number that is one more than **31**.
Underline the number that is one less than **31**.

28 29 31 30 32 27

3. Number the clockface.

Show **1:00** on the clock.

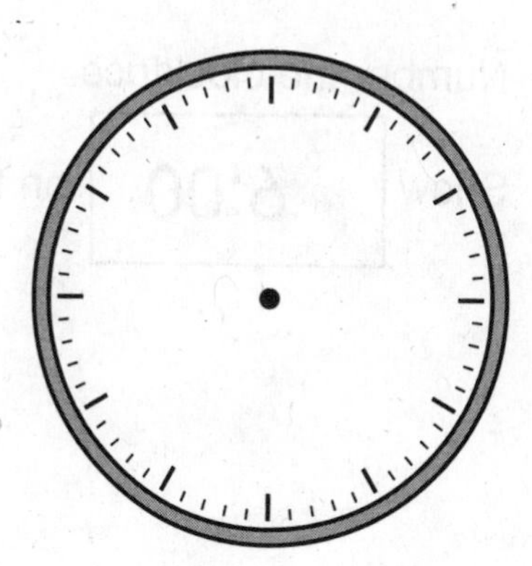

4. Circle each square that has equal-size pieces.

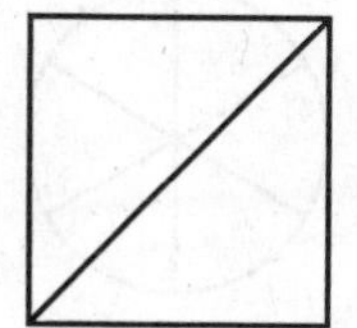 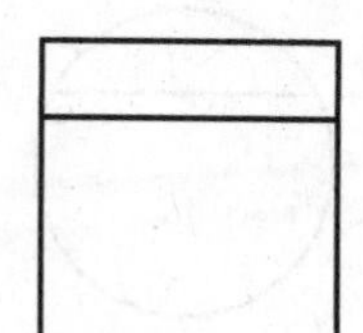 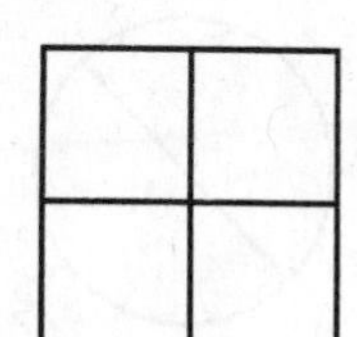 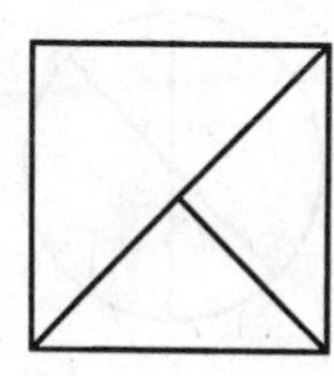

5. Write these numbers.

thirty-seven ______ forty ______ sixteen ______

6. Find the sums.

$6 + 1 =$ ____ $8 + 8 =$ ____ $1 + 8 =$ ____ $6 + 6 =$ ____

Name ______________________________

Date ______________________________

Homework 12B

***Saxon** Math 2 (for use with **Lesson 12**)*

1. There were 10 children on the bus. One child got off the bus. How many children are on the bus now? What type of story problem is this?

Circle one: Some, some more   Some, some went away

2. Circle the number that is one more than **26.**
Underline the number that is one less than **26.**

27   29   26   24   28   25

3. Number the clockface.

Show 6:00 on the clock.

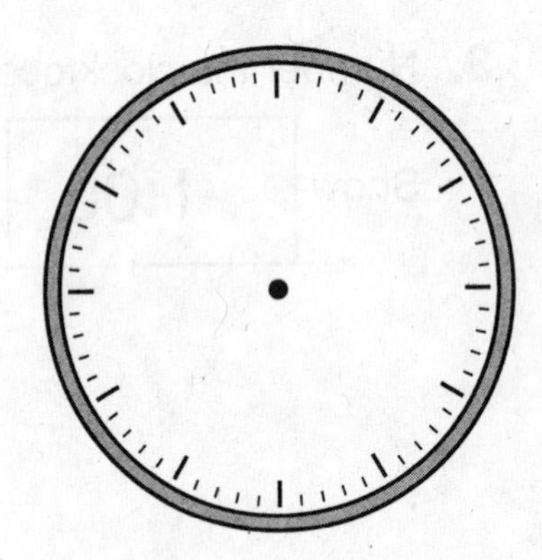

4. Circle each circle that has equal-size pieces.

  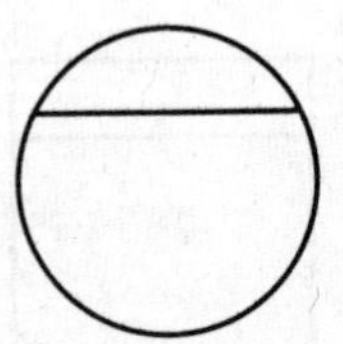 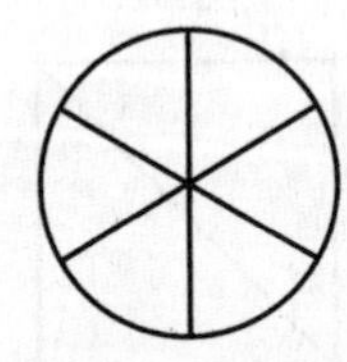

5. Write these numbers.

twenty-seven ______   eighty ______   thirteen ______

6. Find the sums.

$7 + 7 =$ ______ $1 + 7 =$ ______ $9 + 9 =$ ______ $5 + 1 =$ ______

Name ___________________________

***Saxon** Math 2 (for use with **Lesson 13**)*

Set 2: Adding 1 and 0; Doubles

A.

$1 + 0 =$ ____	$3 + 1 =$ ____	$1 + 7 =$ ____
$0 + 4 =$ ____	$1 + 9 =$ ____	$7 + 0 =$ ____
$8 + 1 =$ ____	$0 + 3 =$ ____	$1 + 2 =$ ____
$6 + 0 =$ ____	$4 + 1 =$ ____	$0 + 9 =$ ____
$1 + 6 =$ ____	$5 + 0 =$ ____	$9 + 1 =$ ____
$0 + 2 =$ ____	$1 + 1 =$ ____	$8 + 0 =$ ____
$5 + 1 =$ ____	$0 + 7 =$ ____	$0 + 0 =$ ____

B.

$3 + \square = 3$	$6 + \square = 7$	$1 + \square = 9$
$\square + 8 = 8$	$\square + 3 = 4$	$\square + 1 = 6$

Name ______________________

Fact Homework 13B

Saxon Math 2 (for use with Lesson 13)

Set 2: Adding 1 and 0; Doubles

A. 1. Read the answers to someone.
2. Write the answers.
3. Ask someone to correct your paper. Corrected by ______________

1 + 9 = ____	0 + 5 = ____	6 + 1 = ____
9 + 0 = ____	1 + 4 = ____	0 + 6 = ____
2 + 1 = ____	0 + 7 = ____	1 + 8 = ____
3 + 0 = ____	9 + 1 = ____	4 + 0 = ____
7 + 7 = ____	0 + 1 = ____	1 + 3 = ____
0 + 8 = ____	1 + 1 = ____	2 + 0 = ____
1 + 5 = ____	7 + 0 = ____	0 + 0 = ____

B. Fill in the missing numbers.

5 + ☐ = 5	2 + ☐ = 3	1 + ☐ = 7
☐ + 2 = 2	☐ + 5 = 6	☐ + 1 = 8

Name ______________________________

1	2	3	4	5	6	7	8	9	10
11	12	13	14	15	16	17	18	19	20
21	22	23	24	25	26	27	28	29	30
31	32	33	34	35	36	37	38	39	40
41	42	43	44	45	46	47	48	49	50
51	52	53	54	55	56	57	58	59	60
61	62	63	64	65	66	67	68	69	70
71	72	73	74	75	76	77	78	79	80
81	82	83	84	85	86	87	88	89	90
91	92	93	94	95	96	97	98	99	100

Green—__________ numbers Orange—__________ numbers

Parent: This is a hundred number chart that we will use to examine number patterns. Save it to use at home to help with homework assignments.

Name ______________________________

Date ______________________________

Guided Class Practice 13A

***Saxon** Math 2 (for use with **Lesson 13**)*

1. Use the class birthday graph to answer these questions.

How many children have birthdays in November? ________

What is the third month of the year? ________________

How many children have birthdays in that month? ________

2. Circle the number that is one less than **24**.
Underline the number that is one more than **28.**

25 34 22 27 23 29

3. One of these is my favorite bear.
Use the clues to find my favorite bear.

It is not on the left. Cross out that bear.
It is not in the middle. Cross out that bear.
It is not fourth. Cross out that bear.
It is not second. Cross out that bear.
Circle my favorite bear.

4. Number the clockface.

Show **2:00** on the clock.

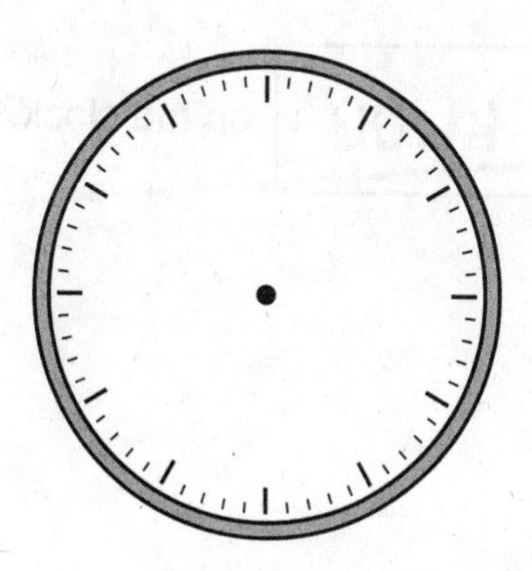

5. Circle all the even numbers.

1, 2, 3, 4, 5, 6, 7, 8, 9, 10,

11, 12, 13, 14, 15, 16, 17, 18, 19, 20

Name ______________________

Date ______________________

Homework 13B

Saxon Math 2 (for use with Lesson 13)

1. Fill in the missing numbers.

21, 22, ____, ____, ____, ____, ____, ____, ____, ____,

____, ____, ____, ____, ____, 36, ____, ____, ____, ____

2. Circle the number that is one less than **37**.
Underline the number that is one more than **32**.

38 33 39 34 31 36

3. One of these is my favorite bear.
Use the clues to find my favorite bear.

It is not on the left. Cross out that bear.
It is not in the middle. Cross out that bear.
It is not second. Cross out that bear.
It is not last. Cross out that bear.
Circle my favorite bear.

4. Number the clockface.

Show **11:00** on the clock.

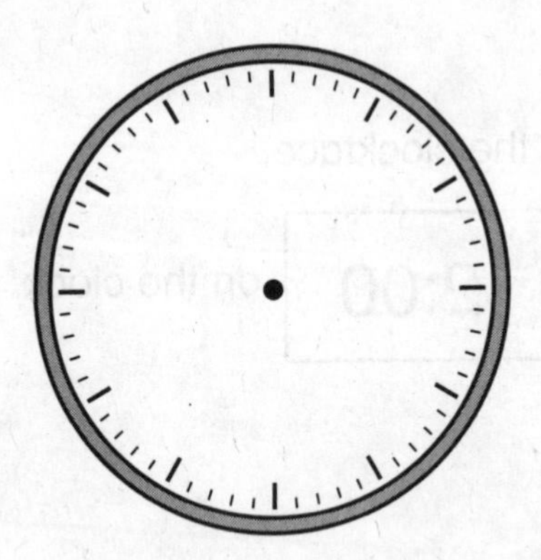

5. Circle all the odd numbers.

1, 2, 3, 4, 5, 6, 7, 8, 9, 10,

11, 12, 13, 14, 15, 16, 17, 18, 19, 20

Name ______________________________

Set 2: Adding 1 and 0; Doubles

9 + 1	3 + 0	4 + 4	7 + 1	0 + 4
1 + 6	8 + 8	9 + 0	3 + 1	0 + 0
8 + 0	1 + 1	0 + 5	7 + 7	2 + 0
5 + 5	4 + 1	1 + 0	6 + 6	8 + 1
2 + 1	7 + 0	1 + 5	0 + 6	9 + 9

Name ____________________

Saxon Math 2 (for use with **Lesson 14**)

Set 2: Adding 1 and 0; Doubles

1. Read the answers to someone.
2. Write the answers.
3. Ask someone to correct your paper. Corrected by ____________________

$1 + 9 =$	$0 + 1 =$	$7 + 7 =$	$5 + 0 =$	$8 + 1 =$
$6 + 1 =$	$4 + 0 =$	$9 + 9 =$	$1 + 3 =$	$0 + 9 =$
$0 + 0 =$	$1 + 2 =$	$7 + 0 =$	$5 + 1 =$	$4 + 4 =$
$0 + 2 =$	$3 + 3 =$	$7 + 1 =$	$3 + 0 =$	$1 + 1 =$
$8 + 8 =$	$6 + 0 =$	$4 + 1 =$	$0 + 8 =$	$6 + 6 =$

Name ______________________________

Date ______________________________

1. Crystal's dog had seven puppies. She gave four of the puppies to friends. How many puppies does Crystal have left? What type of story problem is this?

Circle one: Some, some more   Some, some went away

2. Circle the fifth letter.
Circle the twelfth letter.
Circle the eighth letter.
Circle the eleventh letter.
Circle the second letter.

O S P A E W A V O R E N

3. Circle the smallest number.
Put an × on the largest number.

27   15   24

4. Continue these repeating patterns.

□, △, □, △, ____, ____, ____, ____, ____

○, △, □, ○, △, □, ○, ____, ____, ____, ____, ____

5. Find the sums.

4 + 0 = ____   8 + 8 = ____   3 + 1 = ____

6 + 6 = ____   0 + 7 = ____   8 + 0 = ____

1 + 9 = ____   6 + 1 = ____   7 + 7 = ____

Name ______________________

Date ______________________

1. Carol had 17 pennies. She found 2 more pennies on the sidewalk. How many pennies does Carol have now? What type of story problem is this?

Circle one: Some, some more Some, some went away

2. Circle the fifth letter.
Circle the twelfth letter.
Circle the second letter.
Circle the ninth letter.
Circle the sixth letter.

S E N T I G A C H R E T

3. Circle the smallest number.
Put an X on the largest number.

39 34 47

4. Continue these repeating patterns.

△, □, △, □, ____, ____, ____, ____, ____

□, △, ○, □, △, ○, □, ____, ____, ____, ____

5. Find the sums.

5 + 5 = ____ 8 + 1 = ____ 9 + 0 = ____

4 + 1 = ____ 0 + 5 = ____ 4 + 4 = ____

0 + 6 = ____ 9 + 9 = ____ 1 + 5 = ____

Name ____________________ Score ______

Fact Assessment 2

Saxon Math 2 (for use with Lesson 15-1)

Set 2: Adding 1 and 0; Doubles

$5 + 0$	$1 + 6$	$4 + 4$	$1 + 0$	$9 + 1$
$0 + 4$	$9 + 9$	$0 + 8$	$1 + 3$	$0 + 0$
$7 + 1$	$0 + 2$	$1 + 5$	$2 + 2$	$7 + 0$
$1 + 1$	$5 + 5$	$0 + 9$	$2 + 1$	$7 + 7$
$3 + 0$	$6 + 6$	$4 + 1$	$6 + 0$	$1 + 8$

Name ______________________________

Lesson Worksheet **15-1A**

***Saxon** Math 2 (for use with **Lesson 15-1**)*

A.

$0 + 2$	$2 + 2$	$4 + 2$	$6 + 2$	$8 + 2$
$2 + 0$	$2 + 2$	$2 + 4$	$2 + 6$	$2 + 8$

B.

$1 + 2$	$3 + 2$	$5 + 2$	$7 + 2$	$9 + 2$
$2 + 1$	$2 + 3$	$2 + 5$	$2 + 7$	$2 + 9$

C.

$0 + 2$	$1 + 2$	$2 + 2$	$3 + 2$	$4 + 2$
$5 + 2$	$6 + 2$	$7 + 2$	$8 + 2$	$9 + 2$

Dear Parent,

Today we practiced the adding 2 facts. Your child will be tested next week on the new addition facts listed below.

$3 + 2$	$4 + 2$	$5 + 2$	$6 + 2$	$7 + 2$	$8 + 2$	$9 + 2$
$2 + 3$	$2 + 4$	$2 + 5$	$2 + 6$	$2 + 7$	$2 + 8$	$2 + 9$

During the next four days, your child will practice these facts in class both orally and in writing. The back of each day's fact sheet will contain the fact practice homework.

Tonight's fact practice is the following:

1. Ask your child to point to the problems in order and to say the answers.
2. Point to the problems in random order and ask your child to say the answers.

Keep this paper at home to use for practice.

Name ______________________________

Date ______________________________

Guided Class Practice **15A**

***Saxon** Math 2 (for use with **Lesson 15-1**)*

1. Fill in the missing days of the week.

Sunday, ______________, ______________, Wednesday,

______________, ______________, Saturday

2. Number the clockface.

It is 9:00 now.

Show the time one hour ago on both clocks.

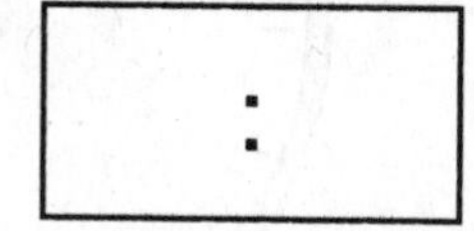

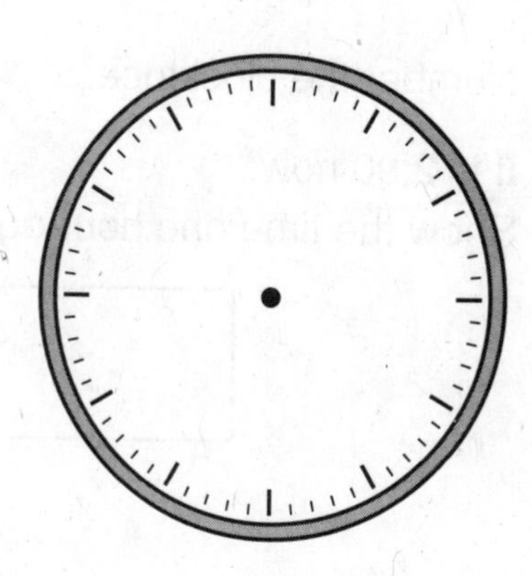

3. What is one more than **39?** ________

What is one less than **20?** ________

4. Write these letters in the squares below.

fourth square **H**
seventh square **S**
eleventh square **N**
ninth square **F**
first square **M**
sixth square **I**
tenth square **U**
third square **T**
second square **A**

Start here*

1	2	3	4	5	6	7	8	9	10	11	12

Name ______________________

Date ______________________

1. Fill in the missing days of the week.

__________, Monday, Tuesday, __________,

Thursday, Friday, __________

2. Number the clockface.

It is 2:00 now.

Show the time one hour ago on both clocks.

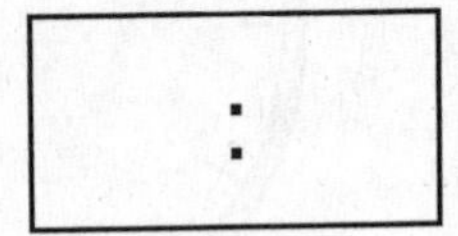

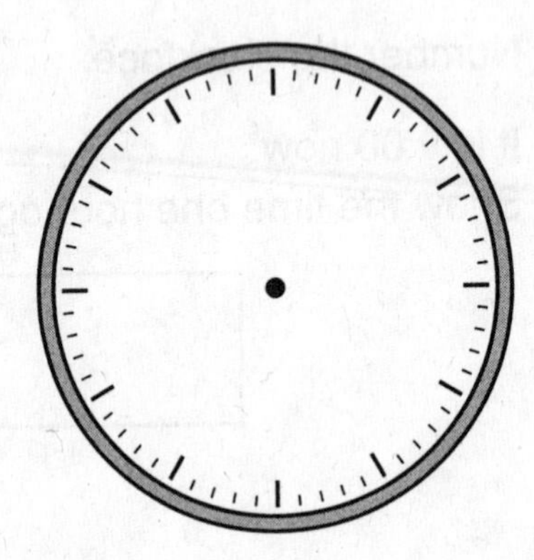

3. What is one more than **49?** ______

What is one less than **40?** ______

4. Write these letters in the squares below.

twelfth square **N**
fourth square **L**
first square **C**
seventh square **E**

eleventh square **O**
ninth square **S**
sixth square **M**

tenth square **O**
third square **L**
second square **A**

Start here*

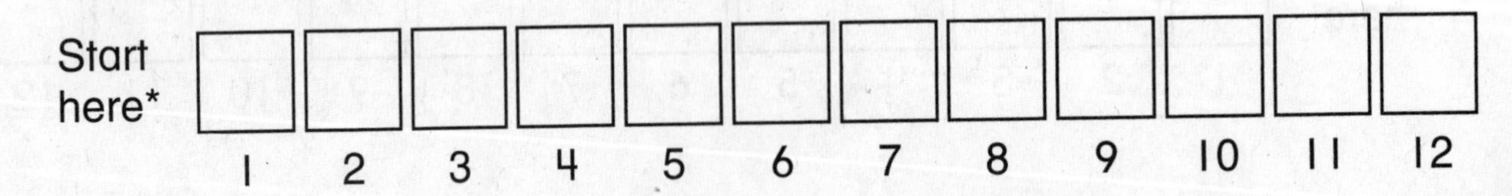

Name ____________________

Date ____________________

Written Assessment 2

***Saxon** Math 2 (for use with **Lesson 15-2**)*

1. Write a **k** in the fifth square.
Write an **a** in the middle square.
Write a **b** in the first square.
Write a **c** in the fourth square.
Write an **l** in the second square.

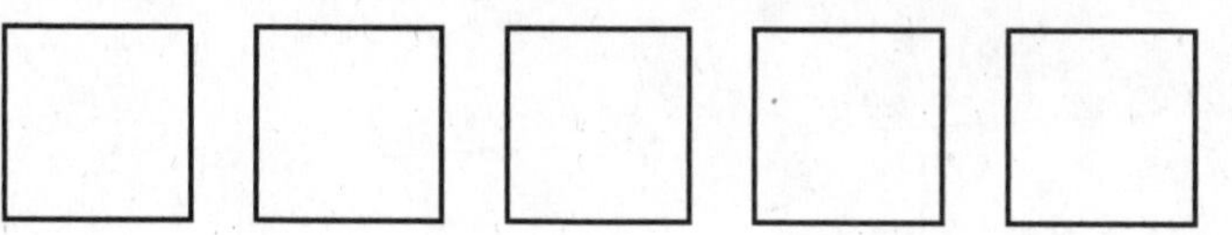

2. Circle the smaller number. 46 42

3. Count by 10's. Fill in the missing numbers.

10, ____, ____, ____, ____, ____, ____, ____, ____, ____

4. Fill in the missing numbers on the chart.

Which number did you write in the square with the **A**? ____

Which number did you write in the square with the **B**? ____

Which number did you write in the square with the **C**? ____

1			4						10
	12							A	
						B			30
		33		C					

5. Draw a square to the right of the triangle.
Draw a circle to the left of the triangle.

6. Find the sums.

8 + 1 = ____ 3 + 1 = ____ 6 + 1 = ____ 7 + 1 = ____

Name ______________________________

Class Fact Practice 16A

***Saxon** Math 2 (for use with **Lesson 16**)*

Set 3: Adding 2; Review Facts

$6 + 2$	$2 + 2$	$8 + 2$	$0 + 2$	$4 + 2$
$2 + 2$	$2 + 6$	$2 + 0$	$2 + 4$	$2 + 8$
$5 + 2$	$9 + 2$	$1 + 2$	$7 + 2$	$3 + 2$
$2 + 1$	$2 + 7$	$2 + 9$	$2 + 3$	$2 + 5$
$6 + 2$	$3 + 2$	$8 + 2$	$5 + 2$	$9 + 2$

Name ____________________

Set 3: Adding 2; Review Facts

1. Read the answers to someone.
2. Write the answers.
3. Ask someone to correct your paper. Corrected by ____________________

$4 + 2$	$0 + 2$	$8 + 2$	$2 + 2$	$6 + 2$
$2 + 8$	$4 + 2$	$2 + 0$	$2 + 6$	$2 + 2$
$3 + 2$	$7 + 2$	$1 + 2$	$9 + 2$	$5 + 2$
$2 + 7$	$2 + 1$	$2 + 9$	$2 + 5$	$2 + 3$
$9 + 2$	$8 + 2$	$5 + 2$	$6 + 2$	$3 + 2$

Name ______________________

Date ______________________

Guided Class Practice 16A

Saxon *Math 2 (for use with* ***Lesson 16****)*

1. Dave had a set of 10 markers. His mother gave him another set of 10 markers. How many markers does Dave have now? What type of story problem is this?

Circle one: Some, some more Some, some went away

2. Fill in the missing days of the week.

Sunday, ______________, ______________,

______________, Thursday,

______________, ______________

3. Put an × on each circle that has equal-size pieces.

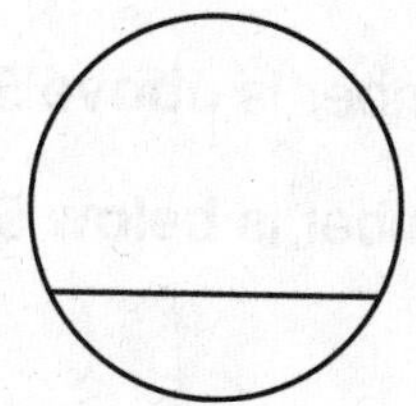

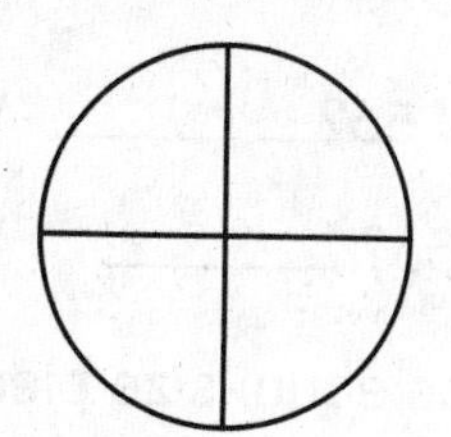

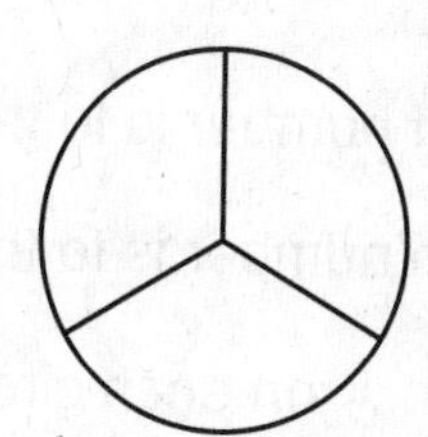

4. Circle the number that is one less than **28.**
Underline the number that is one more than **17.**

18 16 30 29 19 27

5. Write these numbers.

thirty-two ______ fourteen ______

6. Find the sums.
Circle the sums that are even numbers.

6	0	6	4	1	3	9	2
+ 2	+ 7	+ 1	+ 4	+ 8	+ 0	+ 9	+ 5

Name ______________________________

Date ______________________________

Homework 16B

Saxon Math 2 (for use with Lesson 16)

1. Diane had 12 markers. She gave 2 markers to Frank. How many markers does Diane have now? What type of story problem is this?

Circle one: Some, some more Some, some went away

2. Fill in the missing numbers on this piece of a hundred number chart.

41	42	43	44						50
51					56				
									70

Which number is to the **right** of 56? ______ Which number is **above** 56? ______

Which number is to the **left** of 56? ______ Which number is **below** 56? ______

3. Put an X on each circle that has equal-size pieces.

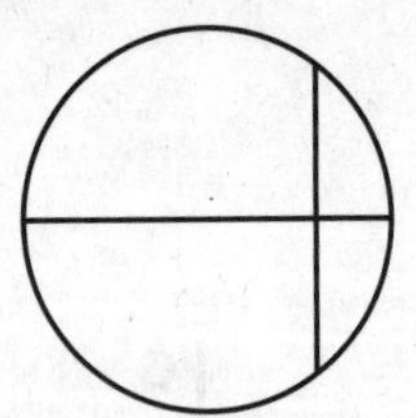
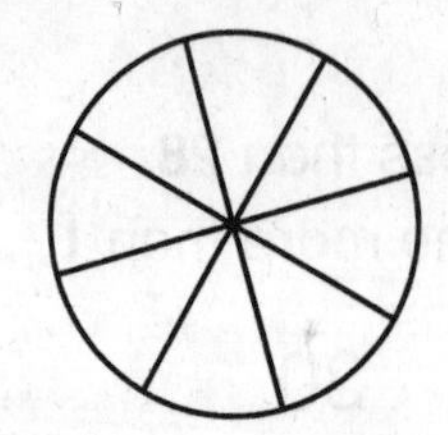
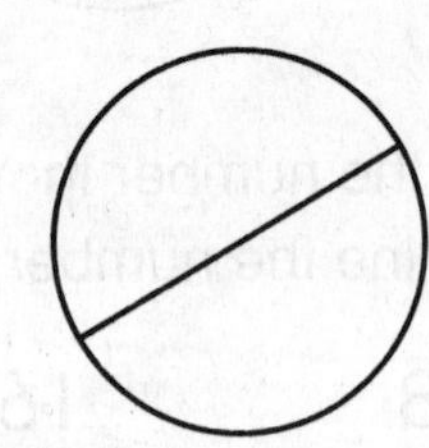

4. Circle the number that is one less than **42.**
Underline the number that is one more than **23.**

41 21 24 43 44 22

5. Write these numbers.

forty-nine ______ seventeen ______

6. Find the sums.
Circle the sums that are even numbers.

$$\begin{array}{r}5\\+\ 0\\\hline\end{array}\quad\begin{array}{r}8\\+\ 8\\\hline\end{array}\quad\begin{array}{r}0\\+\ 9\\\hline\end{array}\quad\begin{array}{r}2\\+\ 3\\\hline\end{array}\quad\begin{array}{r}7\\+\ 1\\\hline\end{array}\quad\begin{array}{r}8\\+\ 2\\\hline\end{array}\quad\begin{array}{r}8\\+\ 0\\\hline\end{array}\quad\begin{array}{r}1\\+\ 4\\\hline\end{array}$$

Name ______________________________

Set 3: Adding 2

$4 + 2$	$1 + 2$	$6 + 2$	$9 + 2$	$5 + 2$
$3 + 2$	$7 + 2$	$0 + 2$	$5 + 2$	$8 + 2$
$2 + 2$	$2 + 9$	$2 + 6$	$2 + 1$	$2 + 4$
$2 + 0$	$2 + 7$	$2 + 3$	$2 + 8$	$2 + 5$
$6 + 2$	$2 + 3$	$9 + 2$	$2 + 4$	$8 + 2$

Name ______________________________

Fact Homework 17B

Saxon *Math 2 (for use with* ***Lesson 17****)*

Set 3: Adding 2

1. Read the answers to someone.
2. Write the answers.
3. Ask someone to correct your paper. Corrected by ______________________

$8 + 2$	$5 + 2$	$0 + 2$	$7 + 2$	$3 + 2$
$2 + 2$	$9 + 2$	$6 + 2$	$1 + 2$	$4 + 2$
$2 + 5$	$2 + 8$	$2 + 3$	$2 + 7$	$2 + 0$
$2 + 4$	$2 + 1$	$2 + 6$	$2 + 9$	$2 + 2$
$8 + 2$	$2 + 4$	$9 + 2$	$2 + 3$	$6 + 2$

Name ______________________________

Our Favorite Days of the Week

Sunday	Monday	Tuesday	Wednesday	Thursday	Friday	Saturday

☺ = 1 child

Name ____________________

Date ____________________

Guided Class Practice 17A

***Saxon** Math 2 (for use with **Lesson 17**)*

1. Use the Favorite Days of the Week class graph to answer the questions.

How many children like Friday best? ______

How many children like Saturday best? ______

Circle the day that more children chose: Friday Saturday

How many children chose the day you circled? ______

2. Fill in the missing days of the week.

____________, Monday, Tuesday, Wednesday,

____________, ____________, ____________

3. Continue the pattern.

△, □, △, ○, △, □, △, ○, △, □, ______, ______, ______, ______

4. Color the sixth shape in Problem 3 red.
Color the eleventh shape in Problem 3 green.
Color the ninth shape in Problem 3 blue.

5. Number the clockface.

It is 4:00 now.
Show the time one hour from now on both clocks.

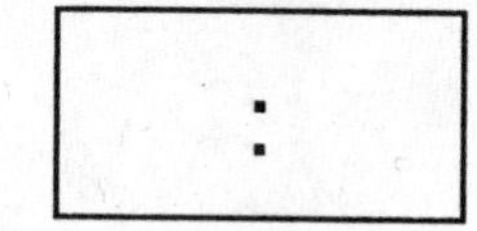

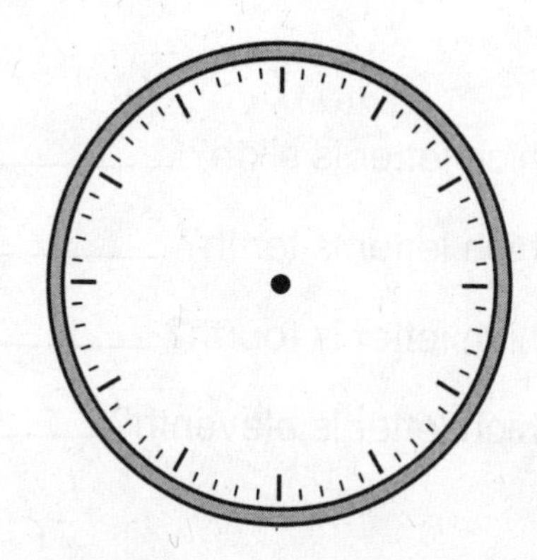

Name ____________________

Date ____________________

1. Fill in the missing days of the week.

Sunday, ____________, ____________, Wednesday,

____________, ____________, Saturday

2. Circle the weekdays in Problem 1.

3. Continue the pattern.

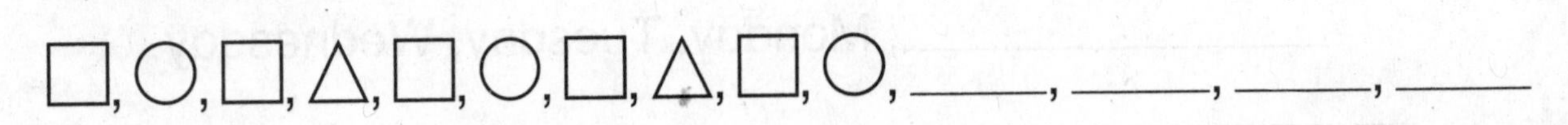

4. Color the seventh shape in Problem 3 orange.
Color the tenth shape in Problem 3 yellow.
Color the fifth shape in Problem 3 brown.

5. Number the clockface.

It is 10:00 now.
Show the time one hour from now on both clocks.

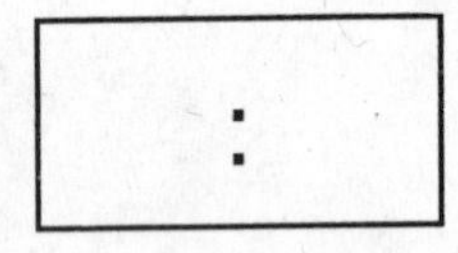

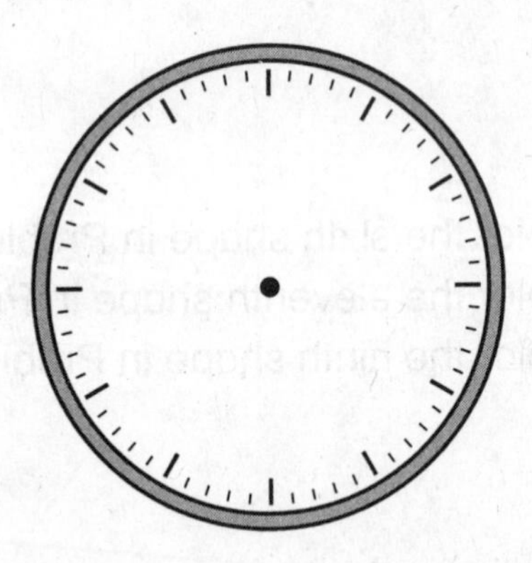

6. Which letter is sixth? ________ M T A L O H B X N E P S

Which letter is tenth? ________

Which letter is fourth? ________

Which letter is eleventh? ________

Name ______________________

Class Fact Practice 18A

***Saxon** Math 2 (for use with **Lesson 18**)*

Set 3: Adding 2

A.

1 + 2 = ____	2 + 4 = ____	7 + 2 = ____
2 + 2 = ____	5 + 2 = ____	2 + 9 = ____
6 + 2 = ____	2 + 0 = ____	3 + 2 = ____
2 + 8 = ____	4 + 2 = ____	2 + 1 = ____
9 + 2 = ____	2 + 5 = ____	8 + 2 = ____
2 + 3 = ____	0 + 2 = ____	2 + 6 = ____
2 + 1 = ____	2 + 7 = ____	2 + 2 = ____

B.

4 + ☐ = 6	2 + ☐ = 10	2 + ☐ = 3
☐ + 2 = 8	☐ + 2 = 9	☐ + 2 = 5

Name ______________________

Saxon Math 2 (for use with Lesson 18)

Set 3: Adding 2

A. 1. Read the answers to someone.
2. Write the answers.
3. Ask someone to correct your paper. Corrected by ____________

2 + 2 = ____	2 + 7 = ____	2 + 1 = ____
2 + 6 = ____	0 + 2 = ____	2 + 3 = ____
8 + 2 = ____	2 + 5 = ____	9 + 2 = ____
2 + 1 = ____	4 + 2 = ____	2 + 8 = ____
3 + 2 = ____	2 + 0 = ____	6 + 2 = ____
2 + 9 = ____	5 + 2 = ____	2 + 2 = ____
7 + 2 = ____	2 + 4 = ____	1 + 2 = ____

B. Fill in the missing numbers.

3 + ☐ = 5	2 + ☐ = 6	2 + ☐ = 4
☐ + 2 = 7	☐ + 2 = 10	☐ + 2 = 2

Name ______________________________

□	△
○	▭

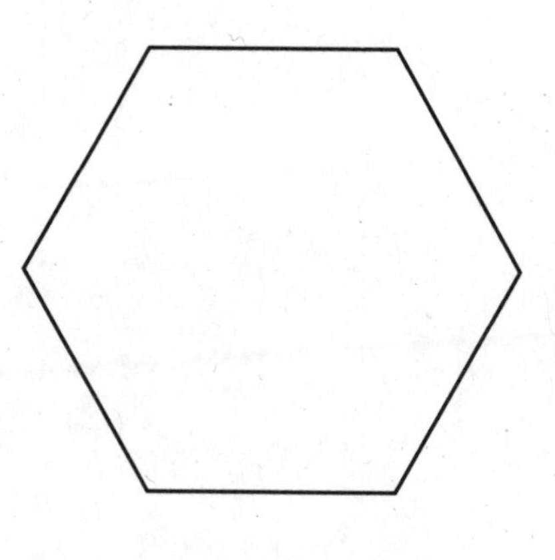

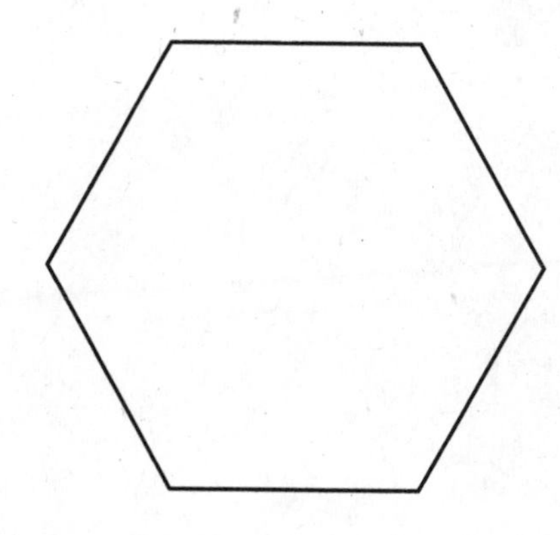

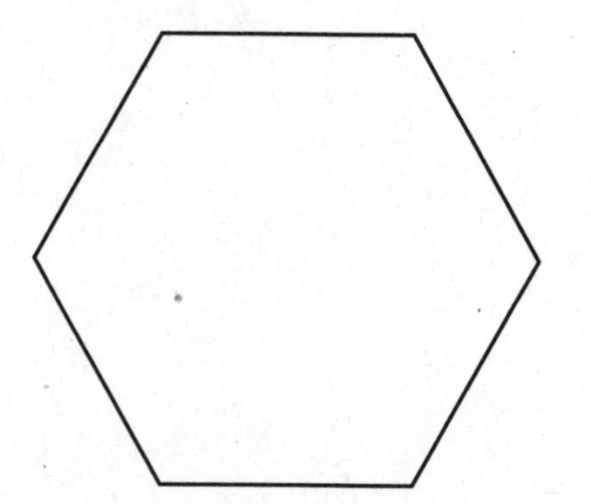

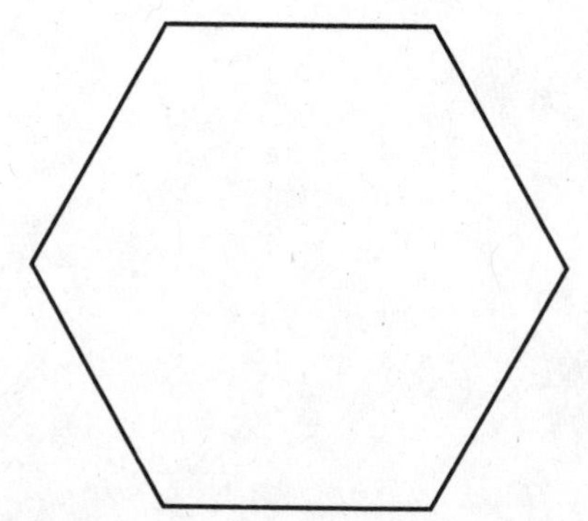

Name ______________________

Date ______________________

Guided Class Practice 18A

***Saxon** Math 2 (for use with **Lesson 18**)*

1. Fran read 6 pages in her book before dinner. After dinner she read 6 more pages. How many pages did she read altogether? What type of story problem is this?

   Circle one: Some, some more Some, some went away

2. Use the Favorite Days of the Week class graph to answer the questions.

   How many children chose Sunday? ______

   How many more children chose Saturday than chose Monday? ______

   How many children chose a weekday? ______

3. Continue the pattern.
   Color the triangles green.
   Color the rectangles orange.
   Color the squares blue.

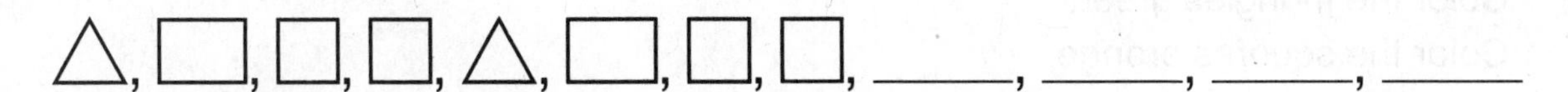

4. Circle the smallest number. 24 35 21

5. Write these numbers. Circle the odd number.

   thirty-one ______ fourteen ______

6. What time does the clock show?

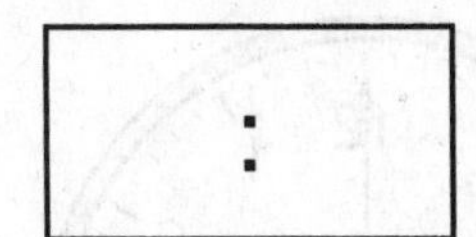

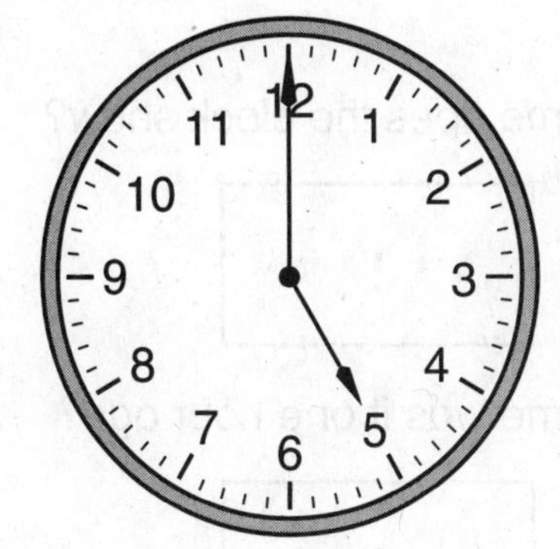

What time was it one hour ago?

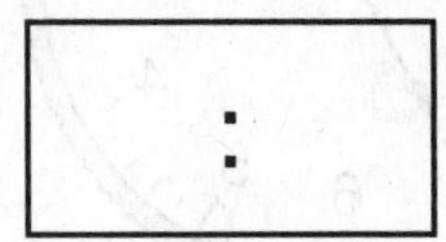

Name ____________________

Date ____________________

Homework 18B

***Saxon** Math 2 (for use with **Lesson 18**)*

1. Mary brought 6 crackers to school. She ate 2 at lunch. How many crackers does she have left to eat? What type of story problem is this?

Circle one: Some, some more Some, some went away

2. Circle the squares that have equal-size pieces.

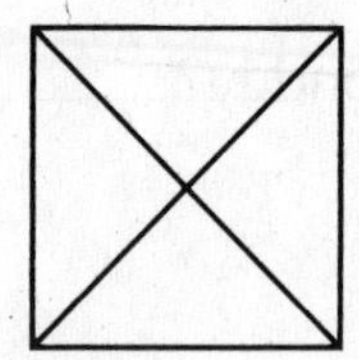
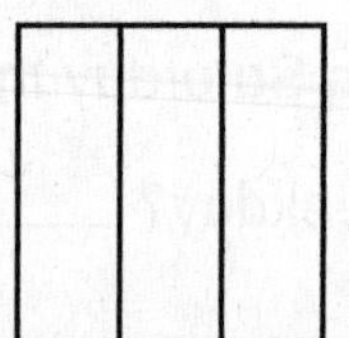
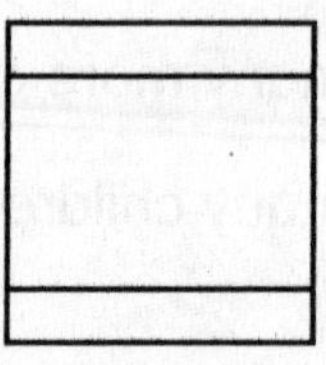

3. Continue the pattern.
Color the circles yellow.
Color the triangles green.
Color the squares orange.

4. Circle the smallest number. 37 21 35

5. Write these numbers. Circle the odd number.

sixteen ______ sixty-one ______

6. What time does the clock show?

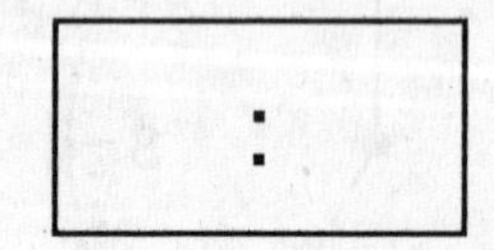

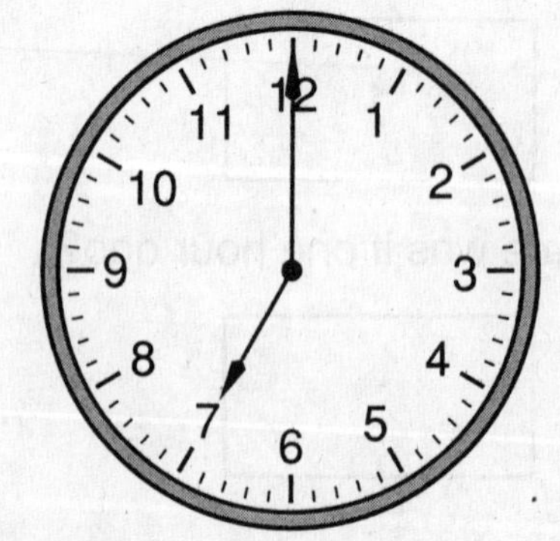

What time was it one hour ago?

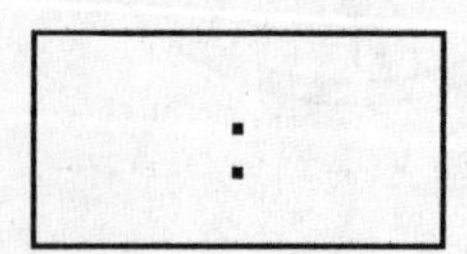

Name ______________________

Saxon Math 2 (for use with Lesson 19)

Set 3: Adding 2; Review Facts

2 + 1	8 + 2	2 + 5	0 + 2	4 + 2
6 + 2	2 + 9	3 + 2	2 + 7	2 + 0
2 + 4	1 + 2	2 + 6	5 + 2	2 + 8
9 + 2	2 + 3	7 + 2	2 + 2	1 + 2
4 + 0	6 + 6	8 + 1	0 + 7	1 + 9

Name ______________________

Fact Homework 19B

***Saxon** Math 2 (for use with **Lesson 19**)*

Set 3: Adding 2; Review Facts

1. Read the answers to someone.
2. Write the answers.
3. Ask someone to correct your paper. Corrected by ______________________

$2 + 8$	$5 + 2$	$2 + 6$	$1 + 2$	$2 + 4$
$2 + 0$	$2 + 7$	$3 + 2$	$2 + 9$	$6 + 2$
$4 + 2$	$0 + 2$	$2 + 5$	$8 + 2$	$2 + 1$
$2 + 2$	$7 + 2$	$2 + 3$	$9 + 2$	$6 + 0$
$7 + 1$	$0 + 3$	$5 + 5$	$1 + 9$	$0 + 0$

Name ______________________________

Date ______________________________

1. The children in Mrs. Hannan's class made this graph.

What day of the week is the children's favorite day? ______________

How many children chose Thursday? ________

How many children are in Mrs. Hannan's class? ________

What 2 days were chosen by the same number of children?

______________ and ______________

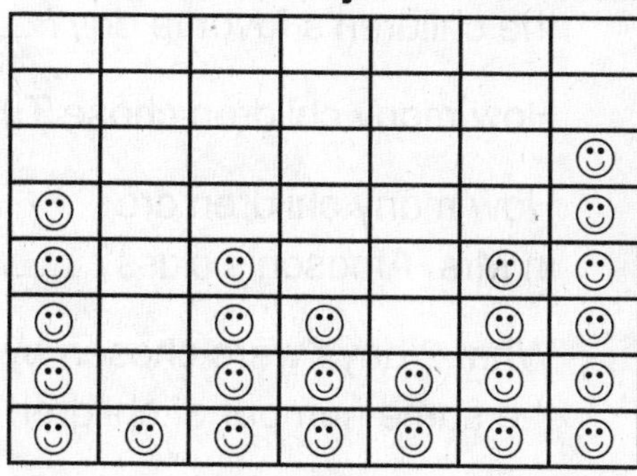

2. Circle the weekdays on the graph in Problem 1.

3. Match each name with the correct picture.

one half •

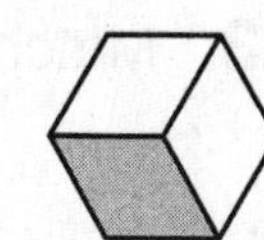

one third •

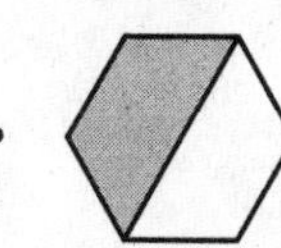

one sixth •

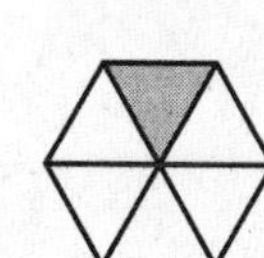

4. Write five odd numbers.

______, ______, ______, ______, ______

5. Color the triangle green.
Color the square orange.
Color the circle blue.
Color the rectangle yellow.

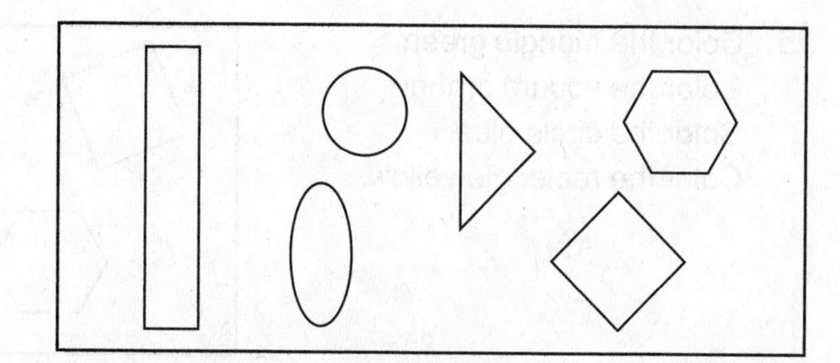

Name ____________________

Date ____________________

1. The children in Mrs. Amason's class made this graph.

What day of the week is the children's favorite day? ____________

How many children chose Tuesday? ________

How many children are in Mrs. Amason's class? ________

What 2 days were chosen by the same number of children?

____________ and ____________

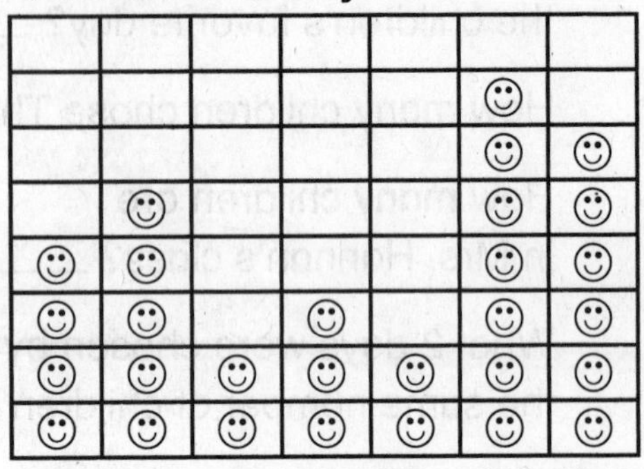

2. Circle the weekend days on the graph in Problem 1.

3. Match each name with the correct picture.

one third •

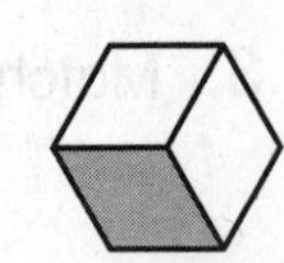

one sixth •

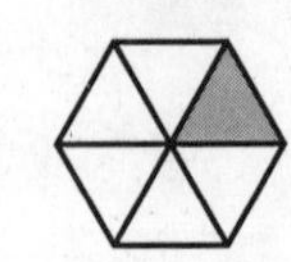

one half •

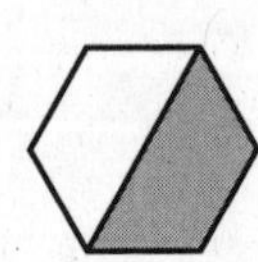

4. Write five even numbers.

______, ______, ______, ______, ______

5. Color the triangle green.
Color the square orange.
Color the circle blue.
Color the rectangle yellow.

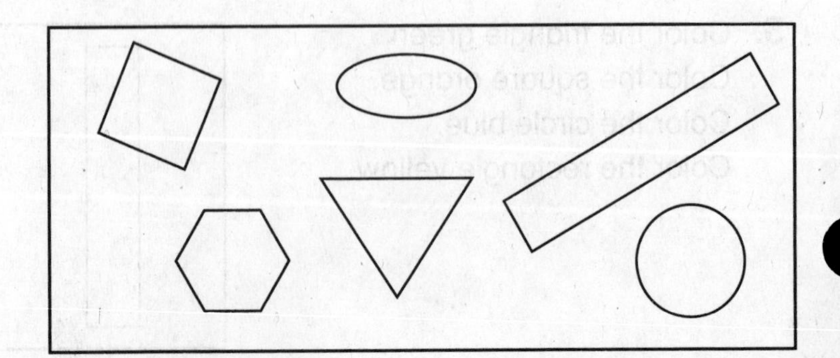

Name ______________________ Score ______

Fact Assessment 3

***Saxon** Math 2 (for use with **Lesson 20-1**)*

Set 3: Adding 2; Review Facts

$1 + 2$	$2 + 7$	$5 + 2$	$2 + 4$	$8 + 2$
$2 + 3$	$0 + 2$	$6 + 2$	$2 + 5$	$2 + 2$
$9 + 2$	$2 + 8$	$3 + 2$	$2 + 1$	$7 + 2$
$2 + 0$	$2 + 9$	$2 + 6$	$4 + 2$	$0 + 0$
$7 + 1$	$0 + 3$	$1 + 9$	$5 + 5$	$6 + 0$

Name ______________________

***Saxon** Math 2 (for use with **Lesson 20-1**)*

A.

$3 + 9 =$ ___ $9 + 3 =$ ___	$4 + 9 =$ ___ $+$ ___	$5 + 9 =$ ___ $+$ ___
$6 + 9 =$ ___ $+$ ___	$7 + 9 =$ ___ $+$ ___	$8 + 9 =$ ___ $+$ ___

B.

$8 + 9$	$9 + 4$	$9 + 2$	$7 + 9$	$9 + 5$
$3 + 9$	$9 + 1$	$9 + 6$	$9 + 9$	$0 + 9$
$5 + 9$	$9 + 8$	$6 + 9$	$9 + 4$	$7 + 9$

Dear Parent,

Today we practiced the adding 9 facts. Your child will be tested next week on the new addition facts listed below:

$3 + 9$	$4 + 9$	$5 + 9$	$6 + 9$	$7 + 9$	$8 + 9$
$9 + 3$	$9 + 4$	$9 + 5$	$9 + 6$	$9 + 7$	$9 + 8$

During the next four days, your child will practice these facts in class both orally and in writing. The back of each day's fact sheet will contain the fact practice homework.

Tonight's fact practice is the following:

1. Ask your child to point to the problems in order and to say the answers.
2. Point to the problems in random order and ask your child to say the answers.

Keep this paper at home to use for practice.

Name ______________________________

Date ______________________________

Problem-Solving Worksheet 20A

***Saxon** Math 2 (for use with **Lesson 20-1**)*

Understand	Plan	Solve	Check

Use Logical Reasoning

Act It Out

John has five hats. The hats are red, blue, green, brown, and orange. John wears a different hat to school each day. John wears the blue hat on the day before the weekend. He wears the brown hat in the middle of the week. John wears the orange hat on the day after he wears the brown hat. John doesn't wear the red hat on Tuesdays. Show which color hat John wears on Mondays.

What color hat does John wear on Mondays? ______________________

Name ______________________

Performance Task Worksheet 20B

***Saxon** Math 2 (for use with **Lesson 20-1**)*

Understand	Plan	Solve	Check

Anjelica has five lunch boxes. The lunch boxes are yellow, red, purple, orange, and blue. She brings a different color lunch box to school each day. She brings the purple lunch box on the fourth school day of the week. She brings the blue lunch box on the second school day of the week. She brings the yellow lunch box on the day after the purple lunch box. She brings the red lunch box on the day after the weekend. Show on which day of the week Anjelica brings the orange lunch box.

On which day of the week does Anjelica bring the orange lunch box?

____

Circle the problem-solving strategies you used to solve this problem.

Act It Out ***Use Logical Reasoning***

Explain how you got your answer: ______________________

Name ______________________________

Date ______________________________

Written Assessment 3

***Saxon** Math 2 (for use with **Lesson 20-2**)*

1. John has 8 pennies. Susan gave him 2 pennies. How many pennies does John have now? What type of story problem is this?

Circle one: Some, some more Some, some went away

Four children were playing. One child went home. How many children are playing now? What type of story problem is this?

Circle one: Some, some more Some, some went away

2. Continue the repeating pattern.

△, ○, ○, □, △, ○, ○, □, ______, ______, ______, ______

3. Write these numbers.

fourteen ______ thirty-five ______

sixty-one ______ seventy ______

4. Circle each shape that has equal-size pieces.

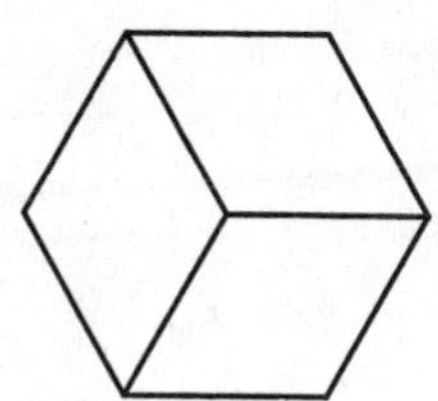 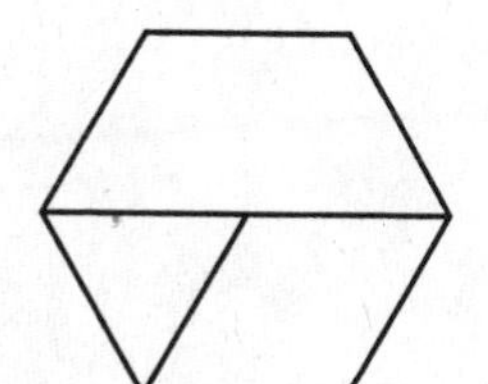 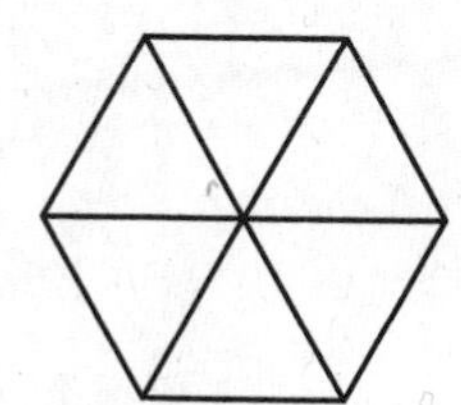

5. What is one more than **49?** ______

What is one less than **20?** ______

6. Add.

$6 + 1 =$ ______	$5 + 0 =$ ______	$8 + 8 =$ ______	$0 + 7 =$ ______
$7 + 7 =$ ______	$1 + 8 =$ ______	$3 + 3 =$ ______	$9 + 9 =$ ______

Name ______________________________

Lesson Worksheet 20-2

***Saxon** Math 2 (for use with **Lesson 20-2**)*

Name ______________________________

Class Fact Practice 21A

***Saxon** Math 2 (for use with **Lesson 21**)*

Set 4: Adding 9

$1 + 9$	$9 + 9$	$5 + 9$	$0 + 9$	$7 + 9$
$4 + 9$	$2 + 9$	$8 + 9$	$3 + 9$	$6 + 9$
$9 + 7$	$9 + 0$	$9 + 9$	$9 + 5$	$9 + 1$
$9 + 3$	$9 + 6$	$9 + 2$	$9 + 8$	$9 + 4$
$5 + 9$	$9 + 7$	$3 + 9$	$9 + 6$	$8 + 9$

Name ______________________

Fact Homework 21B

***Saxon** Math 2 (for use with **Lesson 21**)*

Set 4: Adding 9

1. Read the answers to someone.
2. Write the answers.
3. Ask someone to correct your paper. Corrected by ______________________

$7 + 9$	$0 + 9$	$5 + 9$	$9 + 9$	$1 + 9$
$6 + 9$	$3 + 9$	$8 + 9$	$2 + 9$	$4 + 9$
$9 + 1$	$9 + 5$	$9 + 9$	$9 + 0$	$9 + 7$
$9 + 4$	$9 + 8$	$9 + 2$	$9 + 6$	$9 + 3$
$8 + 9$	$9 + 6$	$3 + 9$	$9 + 7$	$5 + 9$

Name ______________________________

Date ______________________________

Guided Class Practice 21A

***Saxon** Math 2 (for use with **Lesson 21**)*

1. Steven had 10 markers. He gave 3 markers to his friend. How many markers does Steven have now? What type of story problem is this?

Circle one: Some, some more Some, some went away

2. Use the Favorite Days of the Week class graph to answer the questions.

How many children chose a weekday as their favorite day? ______

How many children chose a day of the weekend as their favorite day? ______

3. Match each name with the correct picture.

one half • 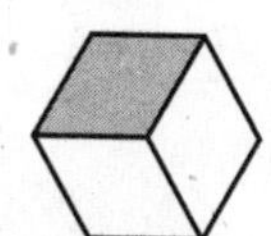

one sixth • 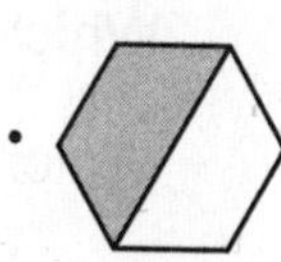

one third • 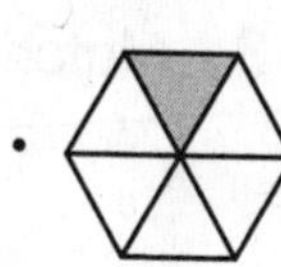

4. Write the names of the first four months.

______________, ______________,

______________, ______________

5. Color the triangle green.
Color the circle yellow.
Color the square orange.
Color the rectangle blue.

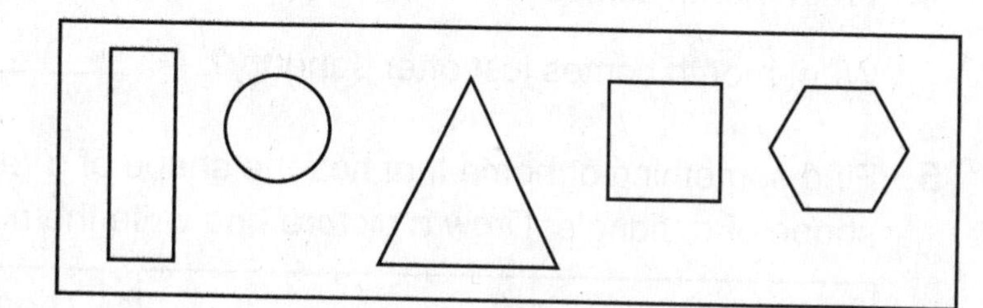

Name ______________________________

Date ______________________________

1. There were 12 eggs in the carton. Scott's family ate 5 eggs for breakfast. How many eggs are left in the carton? What type of story problem is this?

Circle one: Some, some more Some, some went away

2. Fill in the missing numbers on this piece of a hundred number chart.

61	62	63	64						70
71					76				
									90

Which number is to the **right** of 74? ________

Which number is to the **left** of 74? ________

Which number is **above** 74? ________

Which number is **below** 74? ________

3. Match each name with the correct picture.

one half . 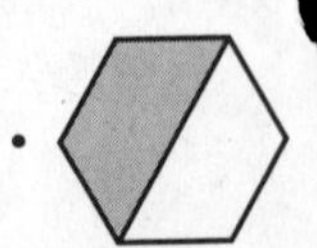

one sixth . 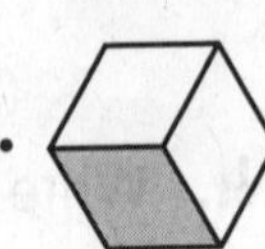

one third . 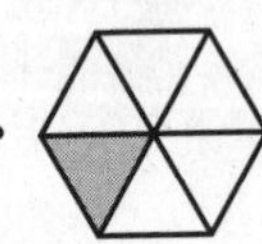

4. What month comes just before April? ____________________

What month comes just after January? ____________________

5. Find something at home that has the shape of a rectangle and something that has the shape of a triangle. Draw a picture and write the name of what you found.

Rectangle	Triangle

Name ______________________________

Class Fact Practice **22A**

***Saxon** Math 2 (for use with **Lesson 22**)*

Set 4: Adding 9

$9 + 2$	$9 + 5$	$9 + 9$	$9 + 7$	$9 + 3$
$6 + 9$	$1 + 9$	$8 + 9$	$0 + 9$	$4 + 9$
$9 + 1$	$9 + 8$	$9 + 6$	$9 + 4$	$9 + 0$
$9 + 9$	$3 + 9$	$2 + 9$	$5 + 9$	$7 + 9$
$9 + 1$	$4 + 9$	$8 + 9$	$9 + 6$	$9 + 2$

Name ______________________________

Fact Homework 22B

***Saxon** Math 2 (for use with **Lesson 22**)*

Set 4: Adding 9

1. Read the answers to someone.
2. Write the answers.
3. Ask someone to correct your paper. Corrected by ______________________

$9 + 0$	$9 + 4$	$9 + 6$	$9 + 8$	$9 + 1$
$7 + 9$	$5 + 9$	$2 + 9$	$3 + 9$	$9 + 9$
$9 + 3$	$9 + 7$	$9 + 9$	$9 + 5$	$9 + 2$
$4 + 9$	$0 + 9$	$8 + 9$	$1 + 9$	$6 + 9$
$9 + 2$	$5 + 9$	$3 + 9$	$9 + 4$	$7 + 9$

Name ______________________________

Date ______________________________

Guided Class Practice 22A

***Saxon** Math 2 (for use with **Lesson 22**)*

1. There were 4 plants in Room 2. The children brought in 3 more plants. How many plants are in Room 2 now? Draw a picture and write a number sentence for this story. Write the answer with a label.

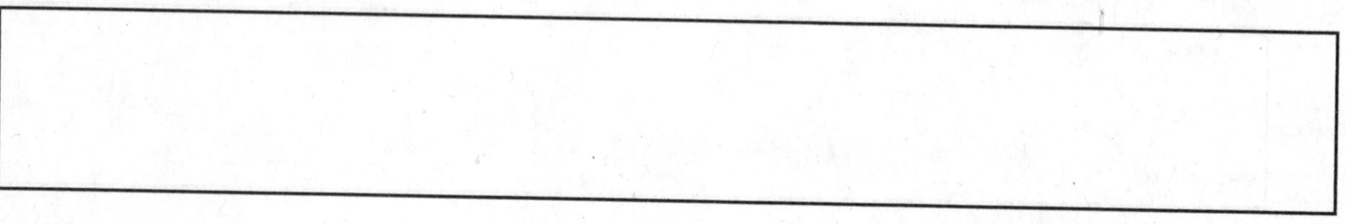

Number sentence ______________________________

Answer ______________________

2. Color one half red. Color one third blue. Color one sixth green.

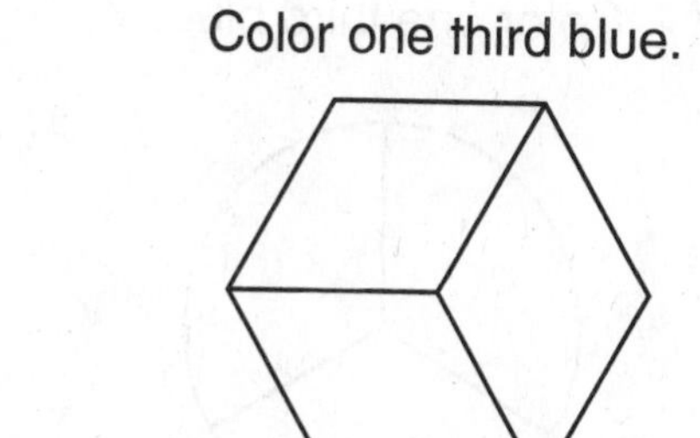

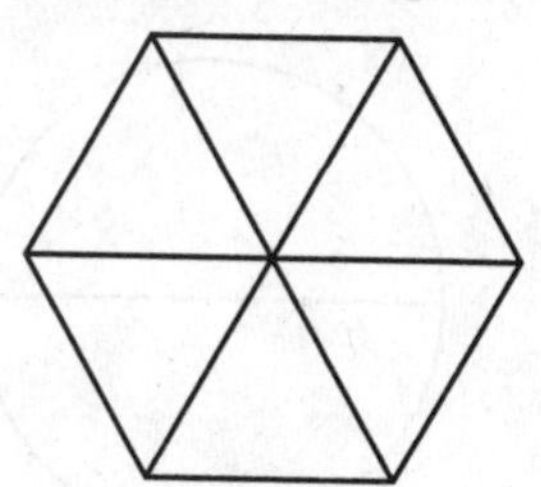

3. Write these numbers.

fifty-nine ______ ninety-five ______

4. Continue the repeating pattern.

Color the pattern: (R = Red, B = Blue, G = Green)

R	B	B	G	R	B	B	G								

5. Find the answers.

$6 + 9 =$ ______ $7 + 9 =$ ______ $4 + 2 =$ ______

6. Color all the squares orange.
Color all the circles yellow.
Color all the triangles green.
Color all the rectangles red.

Name ______________________

Date ______________________

Saxon Math 2 (for use with **Lesson 22**)

1. There were 7 fish in Room 5's fish tank. Mrs. Weber put 2 more fish in the tank. How many fish are in the tank now? Draw a picture and write a number sentence for this story. Write the answer with a label.

Number sentence ______________________

Answer ______________________

2. Color one half red. Color one third blue. Color one sixth green.

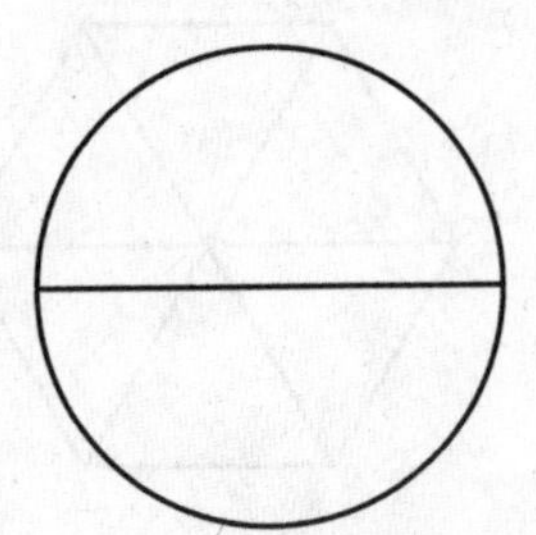

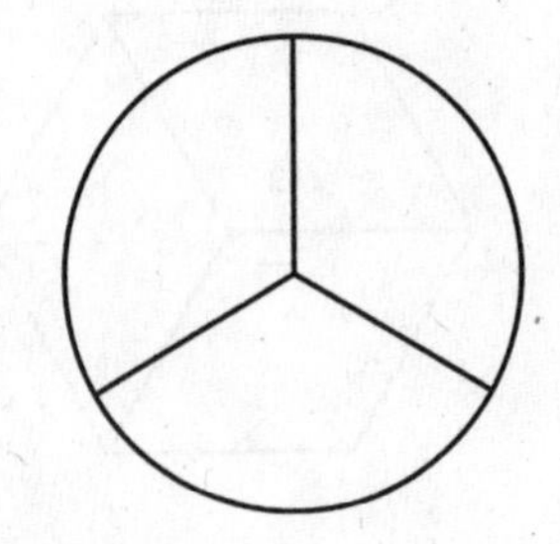

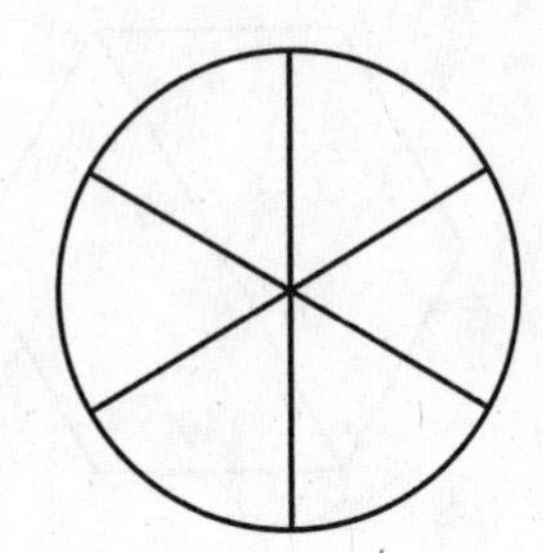

3. Write these numbers.

seventy-eight ______ eighty-seven ______

4. Continue the repeating pattern.

Color the pattern: (R = Red, B = Blue, G = Green)

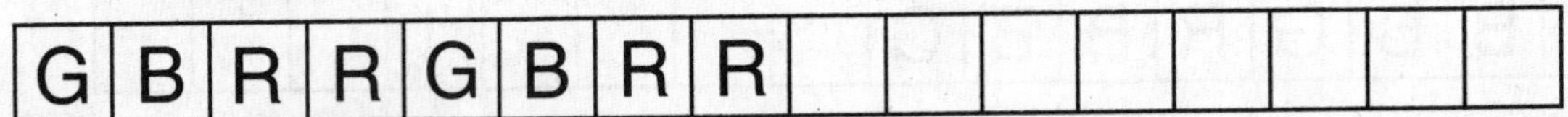

5. Find the answers.

$8 + 9 =$ ______ $5 + 9 =$ ______ $3 + 9 =$ ______

6. Color all the squares orange.
Color all the circles yellow.
Color all the triangles green.
Color all the rectangles red.

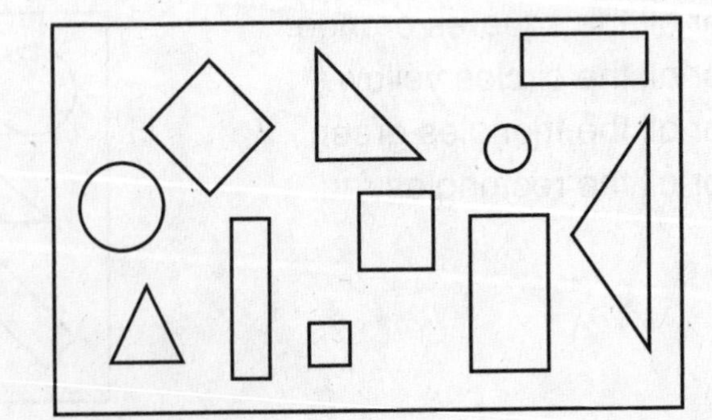

Name ______________________

Set 4: Adding 9

A.

4 + 9 = ____	9 + 2 = ____	0 + 9 = ____
9 + 3 = ____	7 + 9 = ____	9 + 9 = ____
2 + 9 = ____	9 + 6 = ____	5 + 9 = ____
9 + 9 = ____	8 + 9 = ____	9 + 4 = ____
1 + 9 = ____	9 + 7 = ____	6 + 9 = ____
9 + 5 = ____	3 + 9 = ____	9 + 1 = ____
9 + 0 = ____	9 + 8 = ____	4 + 9 = ____

B.

9 + ☐ = 18	9 + ☐ = 15	9 + ☐ = 11
☐ + 9 = 14	☐ + 9 = 12	☐ + 9 = 17

Name ______________________

Set 4: Adding 9

A. 1. Read the answers to someone.
2. Write the answers.
3. Ask someone to correct your paper. Corrected by ______________

4 + 9 = ____	9 + 8 = ____	9 + 0 = ____
9 + 1 = ____	3 + 9 = ____	9 + 5 = ____
6 + 9 = ____	9 + 7 = ____	1 + 9 = ____
9 + 4 = ____	8 + 9 = ____	9 + 9 = ____
5 + 9 = ____	9 + 6 = ____	2 + 9 = ____
9 + 9 = ____	7 + 9 = ____	9 + 3 = ____
0 + 9 = ____	9 + 2 = ____	4 + 9 = ____

B. Fill in the missing numbers.

9 + ☐ = 10	9 + ☐ = 14	9 + ☐ = 17
☐ + 9 = 9	☐ + 9 = 13	☐ + 9 = 16

Name ______________________________

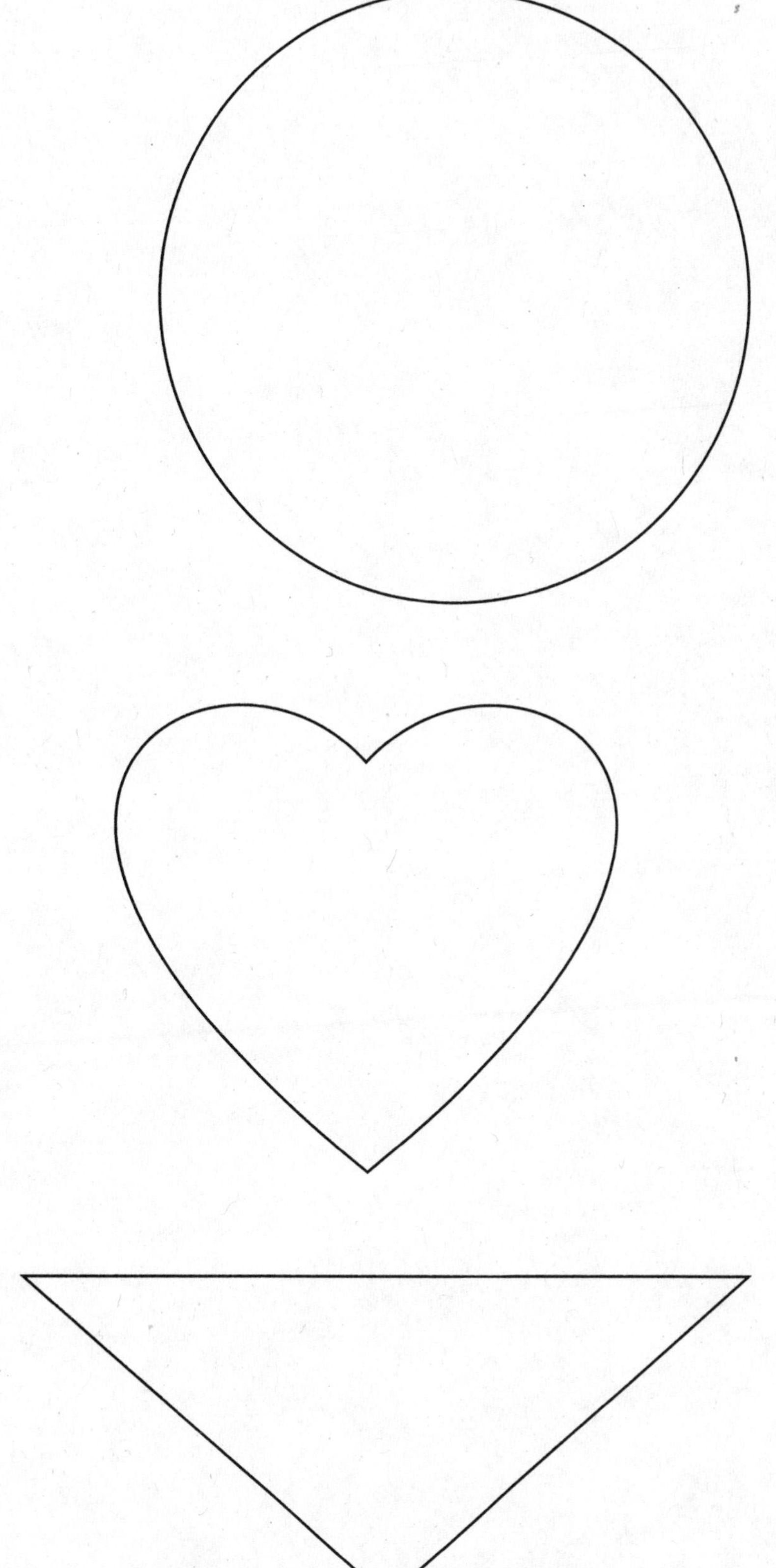

Name ______________________

Date ______________________

Guided Class Practice 23A

***Saxon** Math 2 (for use with **Lesson 23**)*

1. There are 12 girls in Mrs. Miller's class. There are 10 boys in Mrs. Miller's class. How many children are in Mrs. Miller's class? What type of story problem is this?

Circle one: Some, some more Some, some went away

2. Melanie bought five lunch tickets. She used two. How many tickets does she have now? Draw a picture and write a number sentence for this story. Write the answer with a label.

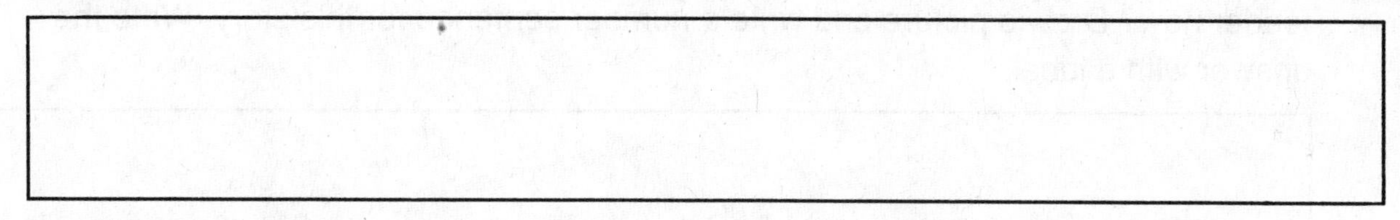

Number sentence ______________________

Answer ______________________

3. Number the clockface.

It is 5:00 now.
Show the time one hour from now on both clocks.

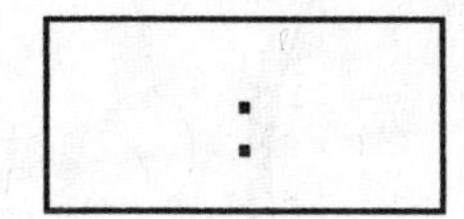

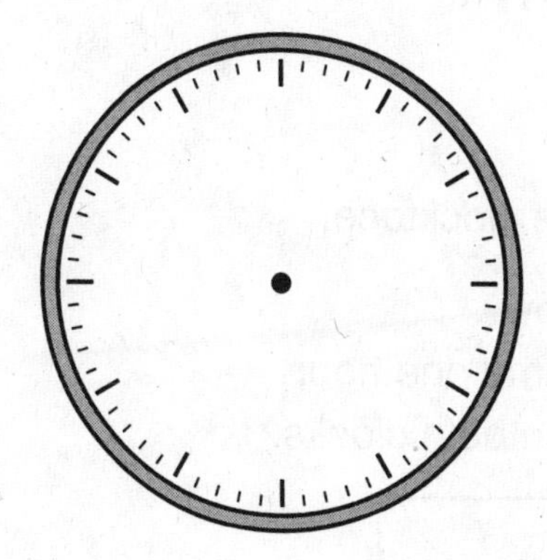

4. Draw a line to divide each shape in half.

Shade one half of each shape.

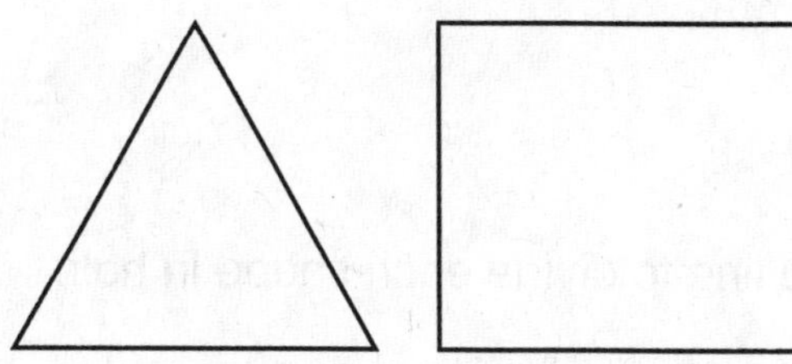

Name ______________________

Date ______________________

Homework 23B

***Saxon** Math 2 (for use with **Lesson 23**)*

1. There are twenty-five children in Mrs. Butler's room. Two of the children went to the gym. How many children are in Mrs. Butler's room now? What type of story problem is this?

Circle one: Some, some more Some, some went away

2. There were five birds at the bird feeder. Three flew away. How many birds are at the feeder now? Draw a picture and write a number sentence for this story. Write the answer with a label.

Number sentence ______________________

Answer ______________________

3. Number the clockface.

It is 6:00 now.
Show the time one hour from now on both clocks.

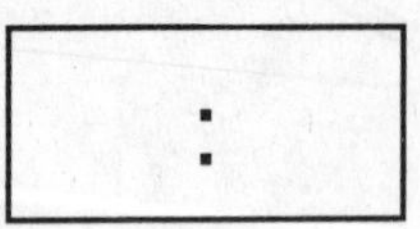

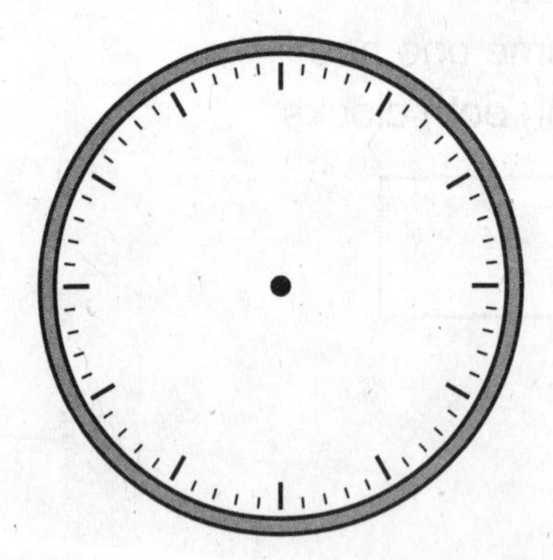

4. Draw a line to divide each shape in half.

Shade one half of each shape.

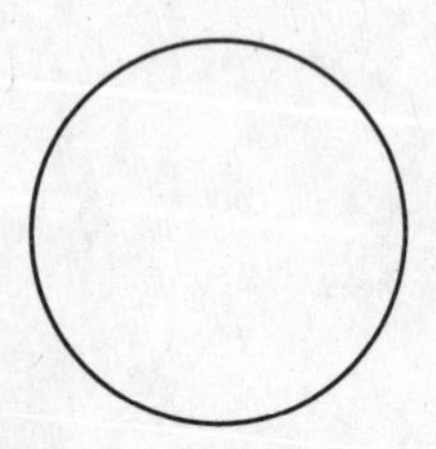

Name ______________________________

Class Fact Practice **24A**

***Saxon** Math 2 (for use with **Lesson 24**)*

Set 4: Adding 9

$6 + 9$	$9 + 3$	$8 + 9$	$4 + 9$	$9 + 1$
$9 + 0$	$5 + 9$	$9 + 8$	$7 + 9$	$9 + 4$
$9 + 6$	$1 + 9$	$3 + 9$	$9 + 5$	$0 + 9$
$9 + 7$	$2 + 9$	$9 + 6$	$8 + 9$	$9 + 9$
$9 + 2$	$9 + 7$	$1 + 9$	$5 + 9$	$9 + 3$

Name ______________________________

Set 4: Adding 9

1. Read the answers to someone.
2. Write the answers.
3. Ask someone to correct your paper. Corrected by ____________________

$0 + 9$	$9 + 5$	$3 + 9$	$1 + 9$	$9 + 6$
$9 + 4$	$7 + 9$	$9 + 8$	$5 + 9$	$9 + 0$
$9 + 1$	$4 + 9$	$8 + 9$	$9 + 3$	$6 + 9$
$9 + 9$	$8 + 9$	$9 + 6$	$2 + 9$	$9 + 7$
$9 + 3$	$9 + 5$	$1 + 9$	$9 + 7$	$9 + 0$

Name ______________________________

Date ______________________________

Guided Class Practice 24A

***Saxon** Math 2 (for use with **Lesson 24**)*

1. Susan has two new pencils and three old pencils. How many pencils does she have? Draw a picture and write a number sentence for this story. Write the answer with a label.

Number sentence ______________________________

Answer ______________________________

2. Write the even numbers from 0 to 18.

___, ___, ___, ___, ___, ___, ___, ___, ___, ___

Write the odd numbers from 1 to 19.

___, ___, ___, ___, ___, ___, ___, ___, ___, ___

3. Circle the greatest number. 47 43 49

4. Show two different ways to divide a square in half.

Shade one half of each square.

5. Number the clockface.

It is 12:00 now.
Show the time one hour from now on both clocks.

:

Name ______________________________

Date ______________________________

Homework 24B

***Saxon** Math 2 (for use with **Lesson 24**)*

1. Sharon's mom bought four blueberry muffins and three apple muffins at the store. How many muffins did she buy? Draw a picture and write a number sentence for this story. Write the answer with a label.

Number sentence ______________________________

Answer ______________________

2. Fill in the missing even numbers from 18 to 0.

18, 16, 14, 12, ____, ____, ____, ____, ____, 0

Fill in the missing odd numbers from 19 to 1.

19, 17, 15, 13, ____, ____, ____, ____, ____, 1

3. Circle the greatest number. 36 34 33

4. Show two different ways to divide a square in half.

Shade one half of each square.

5. Number the clockface.

It is 6:00 now.
Show the time one hour from now on both clocks.

:

Name ______________________ Score ______

Fact Assessment 4

Saxon Math 2 (for use with **Lesson 25-1**)

Set 4: Adding 9

$\begin{array}{r} 2 \\ +\ 9 \\ \hline \end{array}$	$\begin{array}{r} 9 \\ +\ 8 \\ \hline \end{array}$	$\begin{array}{r} 6 \\ +\ 9 \\ \hline \end{array}$	$\begin{array}{r} 9 \\ +\ 1 \\ \hline \end{array}$	$\begin{array}{r} 4 \\ +\ 9 \\ \hline \end{array}$
$\begin{array}{r} 7 \\ +\ 9 \\ \hline \end{array}$	$\begin{array}{r} 9 \\ +\ 3 \\ \hline \end{array}$	$\begin{array}{r} 9 \\ +\ 9 \\ \hline \end{array}$	$\begin{array}{r} 5 \\ +\ 9 \\ \hline \end{array}$	$\begin{array}{r} 8 \\ +\ 9 \\ \hline \end{array}$
$\begin{array}{r} 9 \\ +\ 2 \\ \hline \end{array}$	$\begin{array}{r} 9 \\ +\ 6 \\ \hline \end{array}$	$\begin{array}{r} 9 \\ +\ 4 \\ \hline \end{array}$	$\begin{array}{r} 0 \\ +\ 9 \\ \hline \end{array}$	$\begin{array}{r} 9 \\ +\ 7 \\ \hline \end{array}$
$\begin{array}{r} 9 \\ +\ 5 \\ \hline \end{array}$	$\begin{array}{r} 1 \\ +\ 9 \\ \hline \end{array}$	$\begin{array}{r} 9 \\ +\ 0 \\ \hline \end{array}$	$\begin{array}{r} 3 \\ +\ 9 \\ \hline \end{array}$	$\begin{array}{r} 8 \\ +\ 9 \\ \hline \end{array}$
$\begin{array}{r} 2 \\ +\ 9 \\ \hline \end{array}$	$\begin{array}{r} 4 \\ +\ 9 \\ \hline \end{array}$	$\begin{array}{r} 9 \\ +\ 6 \\ \hline \end{array}$	$\begin{array}{r} 3 \\ +\ 9 \\ \hline \end{array}$	$\begin{array}{r} 9 \\ +\ 7 \\ \hline \end{array}$

Name ____________________

Saxon Math 2 (for use with **Lesson 25-1**)

5	8	2	9	5	7	6	4	7
+ 4	+ 1	+ 7	+ 0	+ 2	+ 8	+ 2	+ 3	+ 7

$$\begin{array}{r} 5 \\ +\,5 \\ \hline \end{array} \quad \begin{array}{r} 5 \\ +\,6 \\ \hline \end{array} \quad \begin{array}{r} 7 \\ +\,7 \\ \hline \end{array} \quad \begin{array}{r} 8 \\ +\,7 \\ \hline \end{array} \quad \begin{array}{r} 4 \\ +\,4 \\ \hline \end{array} \quad \begin{array}{r} 5 \\ +\,4 \\ \hline \end{array}$$

$$\begin{array}{r} 3 \\ +\,3 \\ \hline \end{array} \quad \begin{array}{r} 4 \\ +\,3 \\ \hline \end{array} \quad \begin{array}{r} 6 \\ +\,6 \\ \hline \end{array} \quad \begin{array}{r} 6 \\ +\,7 \\ \hline \end{array} \quad \begin{array}{r} 8 \\ +\,8 \\ \hline \end{array} \quad \begin{array}{r} 8 \\ +\,9 \\ \hline \end{array}$$

$$\begin{array}{r} 2 \\ +\,2 \\ \hline \end{array} \quad \begin{array}{r} 2 \\ +\,3 \\ \hline \end{array} \quad \begin{array}{r} 5 \\ +\,5 \\ \hline \end{array} \quad \begin{array}{r} 6 \\ +\,5 \\ \hline \end{array} \quad \begin{array}{r} 6 \\ +\,6 \\ \hline \end{array} \quad \begin{array}{r} 7 \\ +\,6 \\ \hline \end{array}$$

$$\begin{array}{r} 8 \\ +\,8 \\ \hline \end{array} \quad \begin{array}{r} 9 \\ +\,8 \\ \hline \end{array} \quad \begin{array}{r} 3 \\ +\,3 \\ \hline \end{array} \quad \begin{array}{r} 3 \\ +\,4 \\ \hline \end{array} \quad \begin{array}{r} 7 \\ +\,7 \\ \hline \end{array} \quad \begin{array}{r} 7 \\ +\,8 \\ \hline \end{array}$$

Dear Parent,

Today we practiced the doubles + 1 addition facts. We call these the doubles + 1 facts because we can find these answers easily if we know the answers for the doubles. For example, if we know that 5 + 5 = 10, we can use this fact to help us find the answer for 5 + 6. Six is one more than five, so we can add 5 + 5 + 1 to find the answer for 5 + 6. Your child will be tested next week on the new doubles + 1 addition facts listed below.

$$\begin{array}{r} 3 \\ +\ 4 \\ \hline \end{array} \quad \begin{array}{r} 4 \\ +\ 5 \\ \hline \end{array} \quad \begin{array}{r} 5 \\ +\ 6 \\ \hline \end{array} \quad \begin{array}{r} 6 \\ +\ 7 \\ \hline \end{array} \quad \begin{array}{r} 7 \\ +\ 8 \\ \hline \end{array} \quad \begin{array}{r} 8 \\ +\ 9 \\ \hline \end{array}$$

$$\begin{array}{r} 4 \\ +\ 3 \\ \hline \end{array} \quad \begin{array}{r} 5 \\ +\ 4 \\ \hline \end{array} \quad \begin{array}{r} 6 \\ +\ 5 \\ \hline \end{array} \quad \begin{array}{r} 7 \\ +\ 6 \\ \hline \end{array} \quad \begin{array}{r} 8 \\ +\ 7 \\ \hline \end{array} \quad \begin{array}{r} 9 \\ +\ 8 \\ \hline \end{array}$$

During the next four days, your child will practice these facts in class both orally and in writing. The back of each day's fact sheet will contain the fact practice homework.

Tonight's fact practice is the following:

1. Ask your child to point to the problems in order and to say the answers.
2. Point to the problems in random order and ask your child to say the answers.

Keep this paper at home to use for practice.

Name ______________________

Date ______________________

1. Four girls were playing at the playground. Stephanie joined them. How many girls are playing now? Draw a picture and write a number sentence for this story. Write the answer with a label.

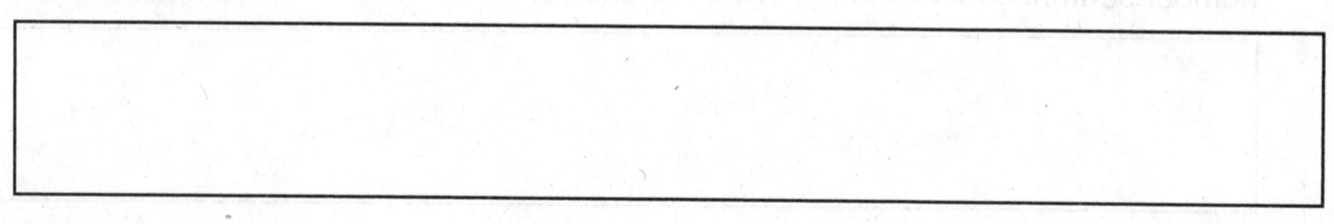

Number sentence ______________________

Answer ______________________

2. Circle the even numbers.

14 7 9 8 4 18 6 3

3. Write the names of the weekdays.

____

____

4. How many children in Rooms 1 and 2 have dogs? ________

Draw pictures to show that 2 children in Room 3 have dogs.

Children with Dogs

Room 1	☺☺☺☺
Room 2	☺☺☺
Room 3	

☺ = 1 child

5. Show two different ways to divide a square in half. Shade one half of each square.

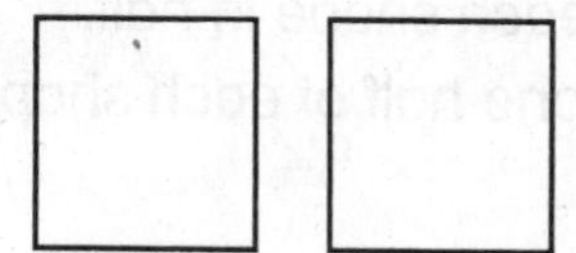

6. Change these doubles facts into doubles plus 1 facts.

$\begin{array}{r} 4 \\ +\ 4 \\ \hline 8 \end{array}$ $\begin{array}{r} \\ +\ \_\_ \end{array}$ $\begin{array}{r} 7 \\ +\ 7 \\ \hline 14 \end{array}$ $\begin{array}{r} \\ +\ \_\_ \end{array}$ $\begin{array}{r} 5 \\ +\ 5 \\ \hline 10 \end{array}$ $\begin{array}{r} \\ +\ \_\_ \end{array}$

Name ______________________

Date ______________________

Homework 25B

***Saxon** Math 2 (for use with **Lesson 25-1**)*

1. At the first bus stop, four children got on the bus. At the next stop, three children got on the bus. How many children are on the bus now? Draw a picture and write a number sentence for this story. Write the answer with a label.

Number sentence ______________________

Answer ______________________

2. Circle the odd numbers.

14 7 9 8 4 18 6 3

3. Circle the names of the days of the weekend.

Sunday, Monday, Tuesday, Wednesday,

Thursday, Friday, Saturday

4. How many children in Rooms 1 and 2 have cats? ________

Draw pictures to show that 4 children in Room 3 have cats.

Children with Cats

Room 1	☺
Room 2	☺☺☺
Room 3	

☺ = 1 child

5. Divide each shape in half. Shade one half of each shape.

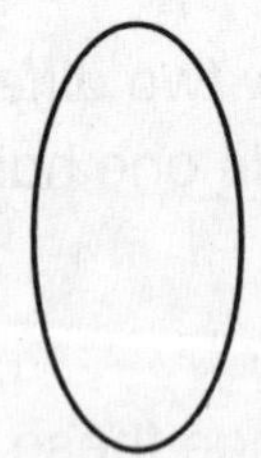

6. Change these doubles facts into doubles plus 1 facts.

$$\begin{array}{r} 3 \\ +\ 3 \\ \hline 6 \end{array} \quad + \_\_\_ \qquad \begin{array}{r} 8 \\ +\ 8 \\ \hline 16 \end{array} \quad + \_\_\_ \qquad \begin{array}{r} 6 \\ +\ 6 \\ \hline 12 \end{array} \quad + \_\_\_$$

Name ______________________

Date ______________________

1. Use the Favorite Days of the Week class graph to answer these questions.

   What day of the week did the most children choose? ____________

   How many children chose the first day of the week as their favorite day? ______

   How many more children chose Saturday than chose Wednesday? ______

2. Write the names of the weekdays. You may use the calendar to help you spell the names of the days.

   ____

   ____

3. Write these letters in the squares below.

   sixth square **E** | tenth square **I** | eighth square **S**
   ninth square **M** | first square **S** | third square **E**
   twelfth square **E** | eleventh square **L** | fifth square **M**
   second square **E**

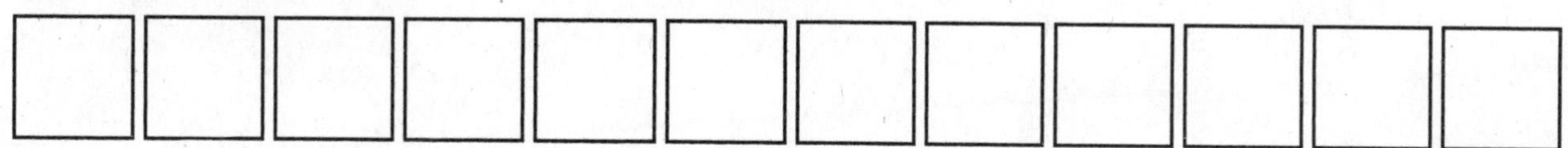

4. Circle the even numbers. 17 6 20 5 14 8

5. Number the clockface.

   It is two o'clock now.
   Show the time one hour from now on both clocks.

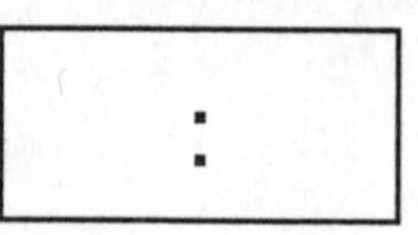

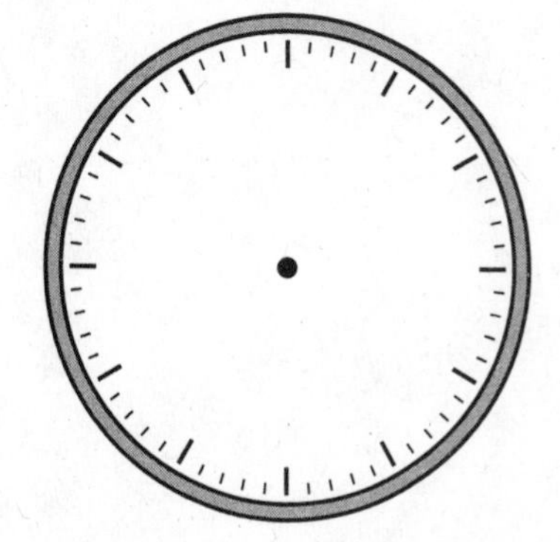

6. Color the triangle green.
   Color the square orange.
   Color the circle yellow.
   Color the rectangle blue.

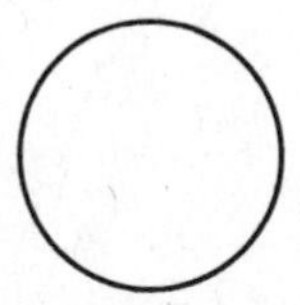

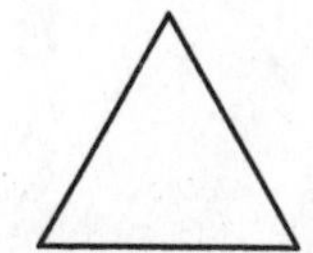

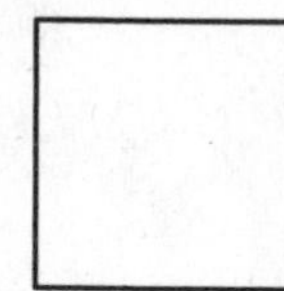

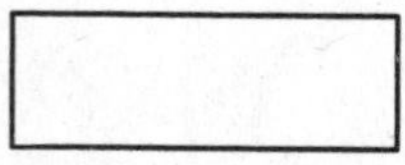

Name ____________________

$5 + 4$	$8 + 1$	$2 + 7$

$9 + 0$	$5 + 2$	$7 + 8$

$6 + 2$	$4 + 3$	$7 + 7$

$5 + 5$ $\quad$ $5 + 6$ $\quad$ $7 + 7$ $\quad$ $8 + 7$ $\quad$ $4 + 4$ $\quad$ $5 + 4$

$3 + 3$ $\quad$ $4 + 3$ $\quad$ $6 + 6$ $\quad$ $6 + 7$ $\quad$ $8 + 8$ $\quad$ $8 + 9$

$2 + 2$ $\quad$ $2 + 3$ $\quad$ $5 + 5$ $\quad$ $6 + 5$ $\quad$ $6 + 6$ $\quad$ $7 + 6$

$8 + 8$ $\quad$ $9 + 8$ $\quad$ $3 + 3$ $\quad$ $3 + 4$ $\quad$ $7 + 7$ $\quad$ $7 + 8$

Dear Parent,

Today we practiced the doubles + 1 addition facts. We call these the doubles + 1 facts because we can find these answers easily if we know the answers for the doubles. For example, if we know that 5 + 5 = 10, we can use this fact to help us find the answer for 5 + 6. Six is one more than five, so we can add 5 + 5 + 1 to find the answer for 5 + 6. Your child will be tested next week on the new doubles + 1 addition facts listed below.

$$\begin{array}{r} 3 \\ +\ 4 \\ \hline \end{array} \quad \begin{array}{r} 4 \\ +\ 5 \\ \hline \end{array} \quad \begin{array}{r} 5 \\ +\ 6 \\ \hline \end{array} \quad \begin{array}{r} 6 \\ +\ 7 \\ \hline \end{array} \quad \begin{array}{r} 7 \\ +\ 8 \\ \hline \end{array} \quad \begin{array}{r} 8 \\ +\ 9 \\ \hline \end{array}$$

$$\begin{array}{r} 4 \\ +\ 3 \\ \hline \end{array} \quad \begin{array}{r} 5 \\ +\ 4 \\ \hline \end{array} \quad \begin{array}{r} 6 \\ +\ 5 \\ \hline \end{array} \quad \begin{array}{r} 7 \\ +\ 6 \\ \hline \end{array} \quad \begin{array}{r} 8 \\ +\ 7 \\ \hline \end{array} \quad \begin{array}{r} 9 \\ +\ 8 \\ \hline \end{array}$$

During the next four days, your child will practice these facts in class both orally and in writing. The back of each day's fact sheet will contain the fact practice homework.

Tonight's fact practice is the following:

1. Ask your child to point to the problems in order and to say the answers.
2. Point to the problems in random order and ask your child to say the answers.

Keep this paper at home to use for practice.

Name ______________________________

Date ______________________________

1. Four girls were playing at the playground. Stephanie joined them. How many girls are playing now? Draw a picture and write a number sentence for this story. Write the answer with a label.

Number sentence ______________________________

Answer ______________________________

2. Circle the even numbers.

14 7 9 8 4 18 6 3

3. Write the names of the weekdays.

____

____

4. How many children in Rooms 1 and 2 have dogs? ________

Draw pictures to show that 2 children in Room 3 have dogs.

Children with Dogs

Room 1	☺☺☺☺
Room 2	☺☺☺
Room 3	

☺ = 1 child

5. Show two different ways to divide a square in half. Shade one half of each square.

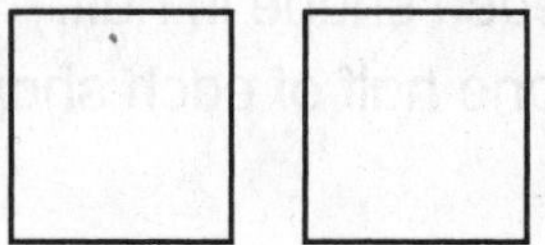

6. Change these doubles facts into doubles plus 1 facts.

$$\begin{array}{r} 4 \\ +\ 4 \\ \hline 8 \end{array} \quad \begin{array}{r} \\ +\ \_\_ \\ \hline \end{array} \qquad \begin{array}{r} 7 \\ +\ 7 \\ \hline 14 \end{array} \quad \begin{array}{r} \\ +\ \_\_ \\ \hline \end{array} \qquad \begin{array}{r} 5 \\ +\ 5 \\ \hline 10 \end{array} \quad \begin{array}{r} \\ +\ \_\_ \\ \hline \end{array}$$

Name ______________________

Date ______________________

Homework 25B

***Saxon** Math 2 (for use with **Lesson 25-1**)*

1. At the first bus stop, four children got on the bus. At the next stop, three children got on the bus. How many children are on the bus now? Draw a picture and write a number sentence for this story. Write the answer with a label.

Number sentence ______________________

Answer ______________

2. Circle the odd numbers.

14 7 9 8 4 18 6 3

3. Circle the names of the days of the weekend.

Sunday, Monday, Tuesday, Wednesday,

Thursday, Friday, Saturday

4. How many children in Rooms 1 and 2 have cats? ______

Draw pictures to show that 4 children in Room 3 have cats.

Children with Cats

Room 1	☺
Room 2	☺☺☺
Room 3	

☺ = 1 child

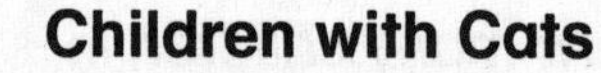

5. Divide each shape in half. Shade one half of each shape.

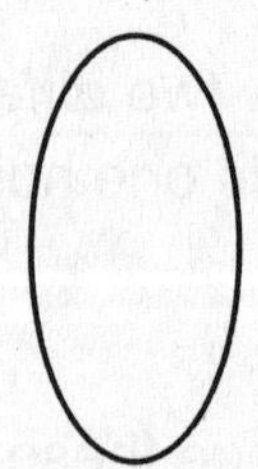

6. Change these doubles facts into doubles plus 1 facts.

$$\begin{array}{r} 3 \\ +\ 3 \\ \hline 6 \end{array} \quad + \underline{\quad} \qquad \begin{array}{r} 8 \\ +\ 8 \\ \hline 16 \end{array} \quad + \underline{\quad} \qquad \begin{array}{r} 6 \\ +\ 6 \\ \hline 12 \end{array} \quad + \underline{\quad}$$

Name ______________________________

Class Fact Practice **26A**

***Saxon** Math 2 (for use with **Lesson 26**)*

Set 5: Doubles + 1

$4 + 3$	$5 + 4$	$6 + 5$	$7 + 6$	$8 + 7$
$9 + 8$	$3 + 4$	$4 + 5$	$5 + 6$	$6 + 7$
$7 + 8$	$8 + 9$	$6 + 5$	$7 + 6$	$4 + 3$
$9 + 8$	$5 + 4$	$8 + 7$	$3 + 4$	$6 + 7$
$5 + 6$	$7 + 8$	$4 + 5$	$8 + 9$	$5 + 6$

Name ______________________________

Set 5: Doubles + 1

1. Read the answers to someone.
2. Write the answers.
3. Ask someone to correct your paper. Corrected by ______________________

$5 + 6$	$8 + 9$	$4 + 5$	$7 + 8$	$3 + 4$
$6 + 7$	$5 + 6$	$8 + 7$	$5 + 4$	$9 + 8$
$4 + 3$	$7 + 6$	$6 + 5$	$8 + 9$	$7 + 8$
$6 + 7$	$5 + 6$	$4 + 5$	$3 + 4$	$9 + 8$
$8 + 7$	$7 + 6$	$6 + 5$	$5 + 4$	$4 + 3$

Name ________________________________

Saxon Math 2 (for use with Lesson 26)

:

:

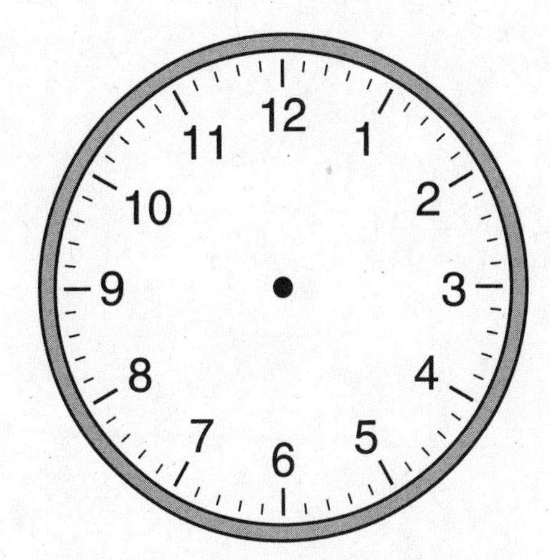

:

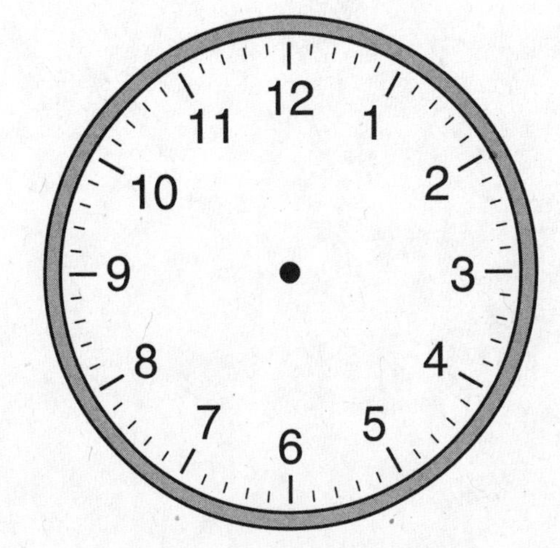

Name ______________________

Date ______________________

1. Last week Leroy ate five peanut butter sandwiches. This week he ate two peanut butter sandwiches and one cheese sandwich. How many peanut butter sandwiches did he eat? Draw a picture and write a number sentence for this story. Write the answer with a label.

Number sentence ______________________

Answer ______________________

2. Number the clockface.

Show half past nine on both clocks.

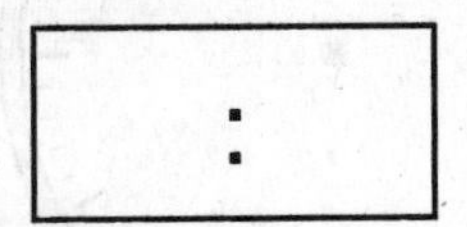

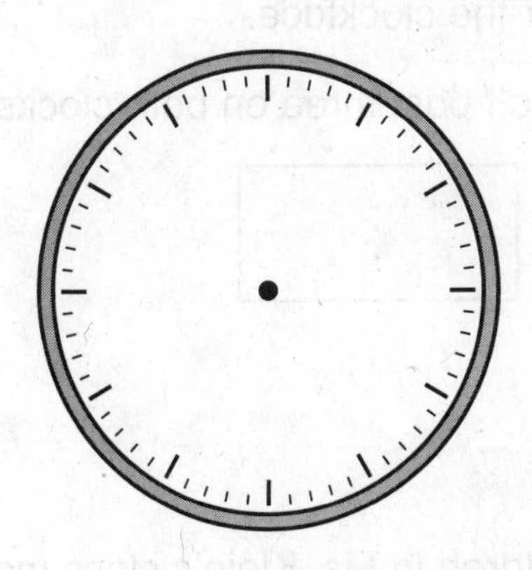

3. The children in Ms. Lee's class made this graph.

What fruit did the fewest children choose? ____________

What fruit did the most children choose? ____________

How many more children chose banana than chose pineapple? ______

Our Favorite Fruits

Apple	☺☺
Orange	☺☺☺
Banana	☺☺☺☺☺☺
Pineapple	☺
Grapes	☺☺☺

☺ = 1 child

4. Write these numbers.

thirty-four ______ forty-three ______

5. Find all the sums. Circle all the doubles plus 1 facts.

$9 + 3 =$ ______ $4 + 5 =$ ______ $6 + 7 =$ ______

$4 + 3 =$ ______ $2 + 7 =$ ______ $9 + 6 =$ ______

Name ______________________________

Date ______________________________

Homework 26B

***Saxon** Math 2 (for use with **Lesson 26**)*

1. There were nine pickles in the jar. Lynn ate three pickles and a bag of chips. How many pickles are left in the jar? Draw a picture and write a number sentence for this story. Write the answer with a label.

Number sentence ______________________________

Answer ______________________

2. Number the clockface.

Show half past three on both clocks.

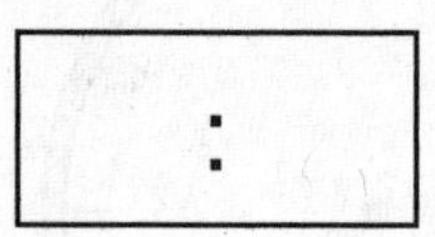

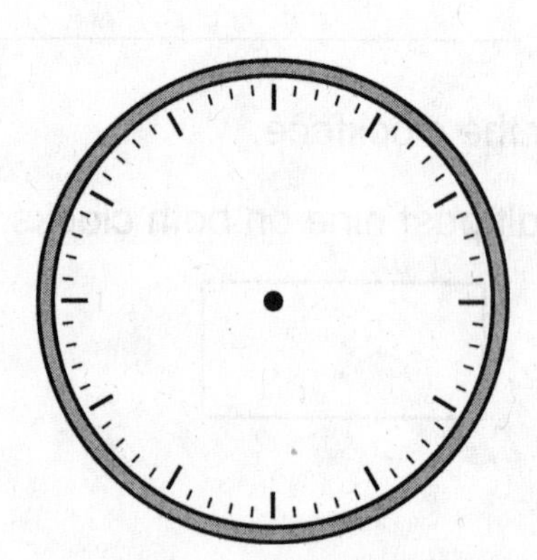

3. The children in Ms. Klein's class made this graph.

What fruit did the fewest children choose? ______________

What fruit did the most children choose? ______________

How many more children chose orange than chose banana? ________

Our Favorite Fruits

Apple	☺☺☺☺
Orange	☺☺☺☺☺
Banana	☺☺☺
Pineapple	☺☺☺☺
Grapes	☺☺

☺ = 1 child

4. Write these numbers.

seventy-two ________ twenty-seven ________

5. Find all the sums. Circle all the doubles plus 1 facts.

$5 + 6 =$ ________ $4 + 9 =$ ________ $8 + 7 =$ ________

$8 + 2 =$ ________ $7 + 6 =$ ________ $9 + 6 =$ ________

Name ______________________________

Set 5: Doubles + 1

3 + 4	6 + 7	8 + 9	4 + 5	7 + 8
5 + 6	4 + 3	9 + 8	7 + 6	5 + 4
8 + 7	6 + 5	6 + 7	4 + 5	5 + 6
3 + 4	8 + 9	7 + 8	5 + 4	7 + 6
9 + 8	4 + 3	6 + 5	8 + 7	5 + 4

Name ______________________________

Fact Homework 27B

***Saxon** Math 2 (for use with **Lesson 27**)*

Set 5: Doubles + 1

1. Read the answers to someone.
2. Write the answers.
3. Ask someone to correct your paper. Corrected by ____________________

$5 + 4$	$8 + 7$	$6 + 5$	$4 + 3$	$9 + 8$
$7 + 6$	$5 + 4$	$7 + 8$	$8 + 9$	$3 + 4$
$5 + 6$	$4 + 5$	$6 + 7$	$6 + 5$	$8 + 7$
$5 + 4$	$7 + 6$	$9 + 8$	$4 + 3$	$5 + 6$
$7 + 8$	$4 + 5$	$8 + 9$	$6 + 7$	$3 + 4$

Name ______________________________

Date ______________________________

Guided Class Practice 27A

***Saxon** Math 2 (for use with **Lesson 27**)*

1. There are four green lunch boxes and three yellow lunch boxes on the shelf in Mrs. Taylor's room. There are two green lunch boxes on the floor. How many green lunch boxes are there altogether? Draw a picture and write a number sentence for this story.

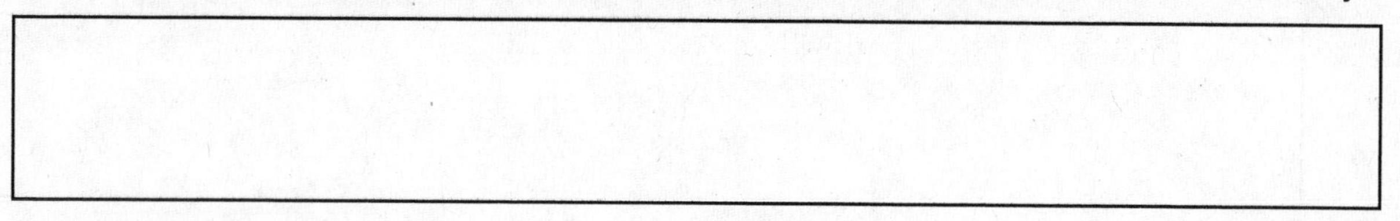

Number sentence ______________________________

Answer ____________________

2. Which number on the thermometer is the temperature closest to? ________°F

°F
110
100
90
80
70
60
50
40
30
20
10
0
−10
−20

3. Continue the repeating pattern.
Color the pattern. (R = Red, G = Green, and Y = Yellow)

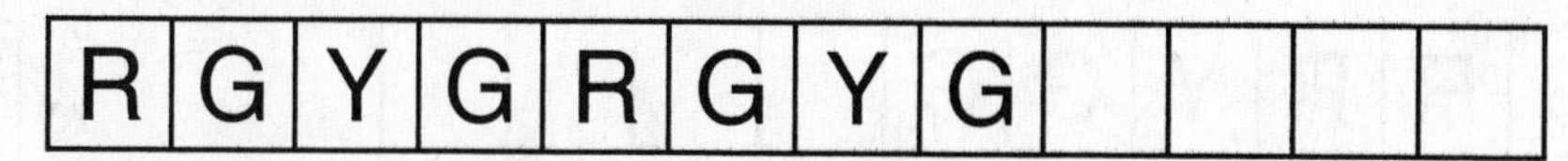

4. Divide the squares in half two different ways.
Color one half of each square red.

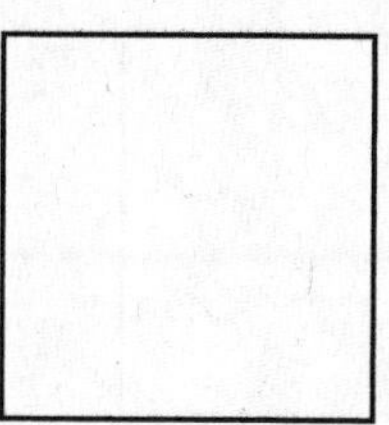

5. Number the clockface.

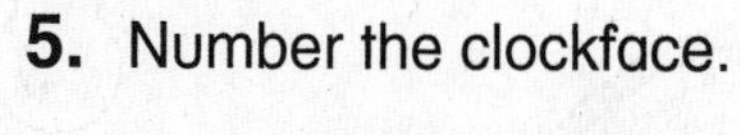

Show half past eight on both clocks.

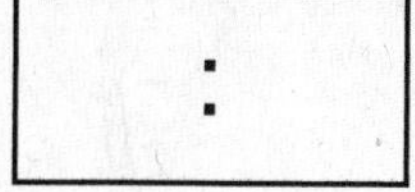

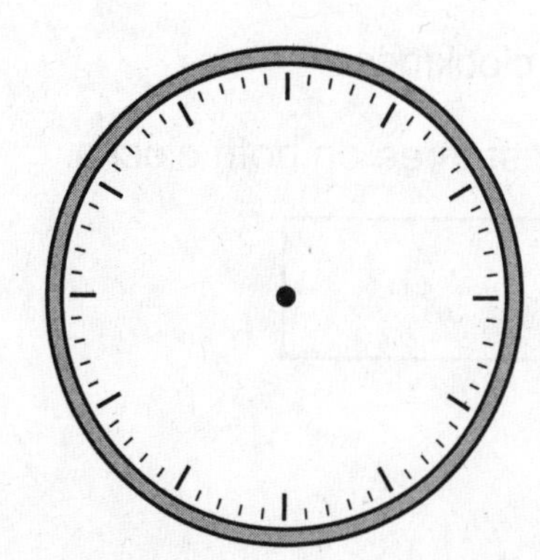

Name ______________________________

Date ______________________________

Homework 27B

***Saxon** Math 2 (for use with **Lesson 27**)*

1. Maureen put four cartons of grape juice and three cartons of apple juice on the table. There are two cartons of apple juice in the refrigerator. How many cartons of apple juice does Maureen have? Draw a picture and write a number sentence for this story.

Number sentence ______________________________

Answer ______________________________

2. Which number on the thermometer is the temperature closest to? ______°F

°F

110
100
90
80
70
60
50
40
30
20
10
0
−10
−20

3. Continue the repeating pattern.
Color the pattern. (R = Red, G = Green, and Y = Yellow)

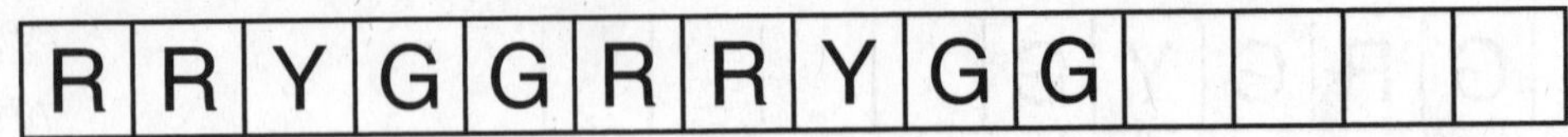

4. Divide the squares in half two different ways.
Color one half of each square red.

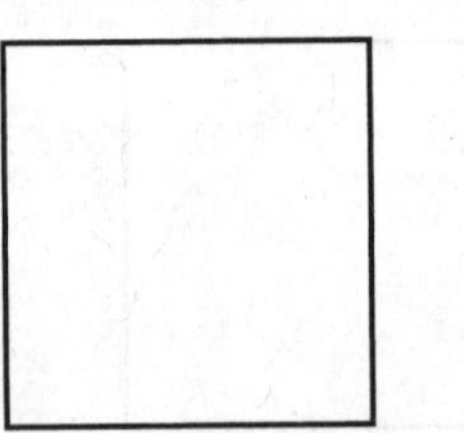

5. Number the clockface.

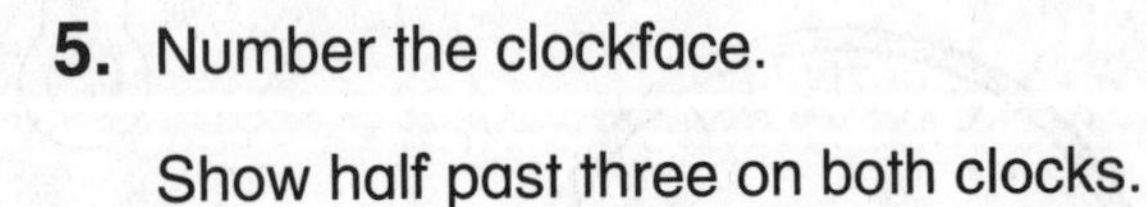

Show half past three on both clocks.

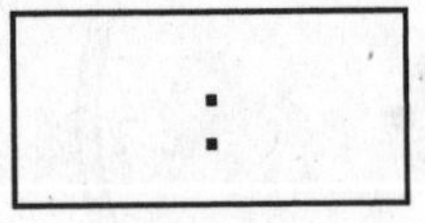

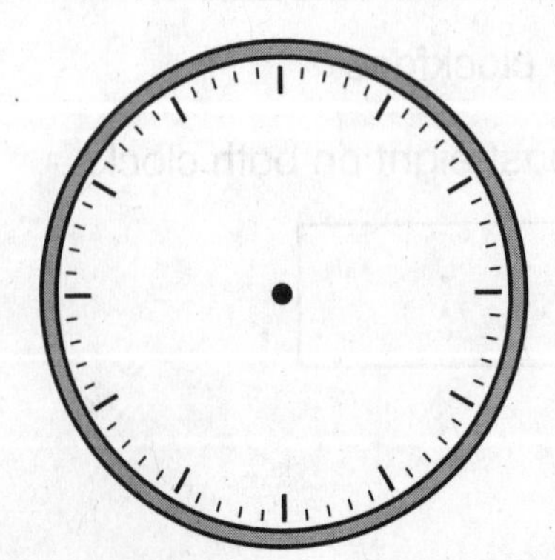

Name ______________________

Set 5: Doubles + 1

A. Write the answers.

Draw lines to connect the problems with the same sum.

3 + 4 = ____ •	• 7 + 6 = ____
7 + 8 = ____ •	• 9 + 8 = ____
6 + 7 = ____ •	• 5 + 4 = ____
4 + 5 = ____ •	• 4 + 3 = ____
8 + 9 = ____ •	• 6 + 5 = ____
5 + 6 = ____ •	• 8 + 7 = ____

B. Fill in the missing addends.

Draw lines to connect the problems that have the same addends.

☐ + 6 = 11 •	• 7 + ☐ = 15
☐ + 8 = 15 •	• 5 + ☐ = 9
☐ + 6 = 13 •	• 5 + ☐ = 11
☐ + 4 = 7 •	• 9 + ☐ = 17
☐ + 4 = 9 •	• 7 + ☐ = 13
☐ + 8 = 17 •	• 3 + ☐ = 7

Name ______________________

Fact Homework 28B

Saxon Math 2 (for use with Lesson 28)

Set 5: Doubles + 1 Corrected by ________

A. Write the answers.

Draw lines to connect the problems with the same sum.

5 + 6 = ____ •	• 5 + 4 = ____
3 + 4 = ____ •	• 7 + 6 = ____
7 + 8 = ____ •	• 6 + 5 = ____
4 + 5 = ____ •	• 9 + 8 = ____
8 + 9 = ____ •	• 4 + 3 = ____
6 + 7 = ____ •	• 8 + 7 = ____

B. Fill in the missing addends.

Draw lines to connect the problems that have the same addends.

☐ + 7 = 13 •	• 8 + ☐ = 17
☐ + 5 = 9 •	• 4 + ☐ = 7
☐ + 9 = 17 •	• 7 + ☐ = 15
☐ + 3 = 7 •	• 6 + ☐ = 11
☐ + 5 = 11 •	• 6 + ☐ = 13
☐ + 8 = 15 •	• 4 + ☐ = 9

Name ______________________

Date ______________________

Guided Class Practice **28A**

Saxon *Math 2 (for use with* ***Lesson 28****)*

1. There were ten apples in a basket. Seth ate three apples. How many apples are in the basket now? Draw a picture and write a number sentence for this story. Write the answer with a label.

Number sentence ______________________

Answer ______________________

2. Which number on the thermometer is the temperature closest to? ________ °F

°F
110
100
90
80
70
60
50
40
30
20
10
0
–10
–20

3. How much money is this? ____________

4. Number the clockface.

Show half past ten.

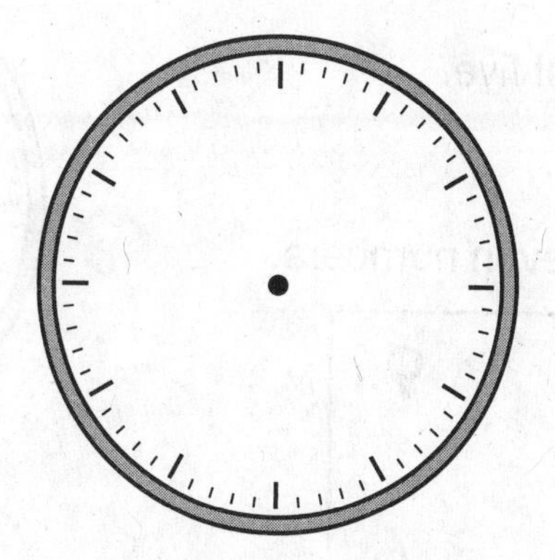

5. Circle all the odd numbers.

15	4	9
18	5	6
3	8	11

Name ______________________________

Date ______________________________

Saxon Math 2 (for use with **Lesson 28**)

1. There were seven flies in the room. Two flew out the window. How many flies are in the room now? Draw a picture and write a number sentence for this story. Write the answer with a label.

Number sentence ______________________________

Answer ____________________

2. Which number on the thermometer is the temperature closest to? ________ °F

°F
110
100
90
80
70
60
50
40
30
20
10
0
–10
–20

3. How much money is this? ____________

4. Number the clockface.

Show half past five.

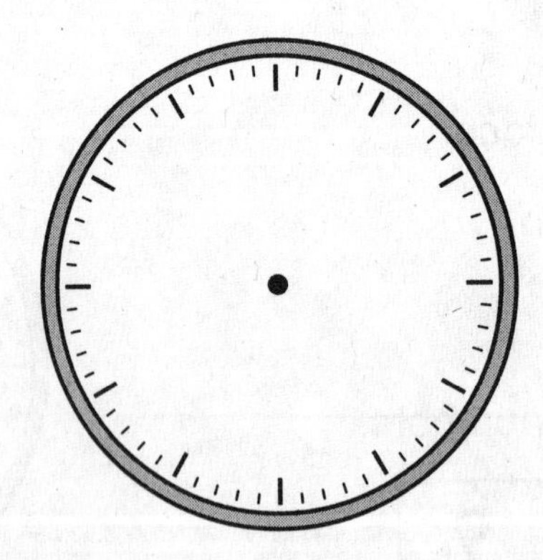

5. Circle all the even numbers.

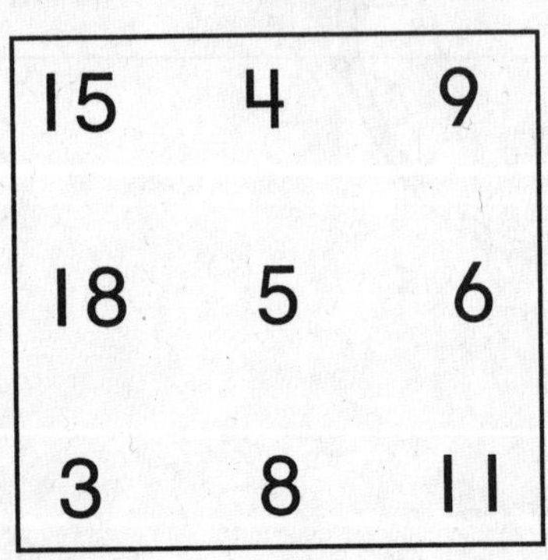

15	4	9
18	5	6
3	8	11

Name ______________________________

Class Fact Practice **29A**

***Saxon** Math 2 (for use with **Lesson 29**)*

Set 5: Doubles + 1

$\begin{array}{r} 3 \\ +\ 4 \\ \hline \end{array}$	$\begin{array}{r} 7 \\ +\ 6 \\ \hline \end{array}$	$\begin{array}{r} 5 \\ +\ 4 \\ \hline \end{array}$	$\begin{array}{r} 8 \\ +\ 9 \\ \hline \end{array}$	$\begin{array}{r} 5 \\ +\ 6 \\ \hline \end{array}$
$\begin{array}{r} 9 \\ +\ 8 \\ \hline \end{array}$	$\begin{array}{r} 4 \\ +\ 5 \\ \hline \end{array}$	$\begin{array}{r} 8 \\ +\ 7 \\ \hline \end{array}$	$\begin{array}{r} 4 \\ +\ 3 \\ \hline \end{array}$	$\begin{array}{r} 6 \\ +\ 7 \\ \hline \end{array}$
$\begin{array}{r} 6 \\ +\ 5 \\ \hline \end{array}$	$\begin{array}{r} 7 \\ +\ 8 \\ \hline \end{array}$	$\begin{array}{r} 4 \\ +\ 5 \\ \hline \end{array}$	$\begin{array}{r} 4 \\ +\ 3 \\ \hline \end{array}$	$\begin{array}{r} 7 \\ +\ 6 \\ \hline \end{array}$
$\begin{array}{r} 5 \\ +\ 6 \\ \hline \end{array}$	$\begin{array}{r} 8 \\ +\ 7 \\ \hline \end{array}$	$\begin{array}{r} 3 \\ +\ 4 \\ \hline \end{array}$	$\begin{array}{r} 9 \\ +\ 8 \\ \hline \end{array}$	$\begin{array}{r} 6 \\ +\ 7 \\ \hline \end{array}$
$\begin{array}{r} 5 \\ +\ 4 \\ \hline \end{array}$	$\begin{array}{r} 7 \\ +\ 8 \\ \hline \end{array}$	$\begin{array}{r} 6 \\ +\ 5 \\ \hline \end{array}$	$\begin{array}{r} 8 \\ +\ 9 \\ \hline \end{array}$	$\begin{array}{r} 3 \\ +\ 4 \\ \hline \end{array}$

Name ______________________________

Saxon Math 2 (for use with **Lesson 29**)

Set 5: Doubles + 1

1. Read the answers to someone.
2. Write the answers.
3. Ask someone to correct your paper. Corrected by ______________________

3 + 4	8 + 9	6 + 5	7 + 8	5 + 4
6 + 7	9 + 8	3 + 4	8 + 7	5 + 6
7 + 6	4 + 3	4 + 5	7 + 8	6 + 5
6 + 7	4 + 3	8 + 7	4 + 5	9 + 8
5 + 6	8 + 9	5 + 4	7 + 6	3 + 4

Name ______________________________

Date ______________________________

Guided Class Practice 29A

***Saxon** Math 2 (for use with **Lesson 29**)*

1. Lee and Gail are having a party. Six friends are coming to the party. How many children will be at the party? Draw a picture and write a number sentence for this story. Write the answer with a label.

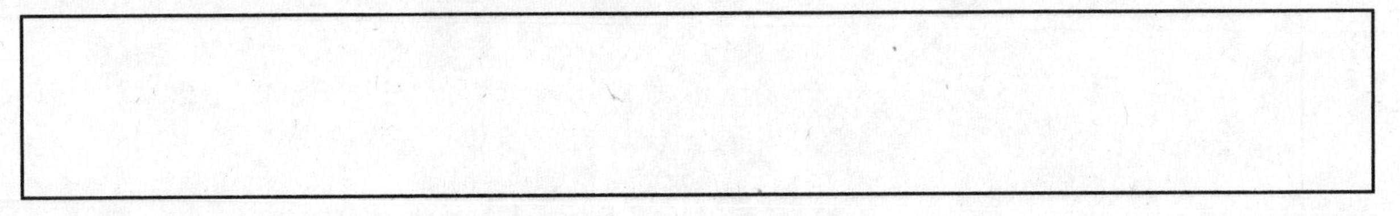

Number sentence ______________________________

Answer ____________________

2. Write an addition and subtraction fact family using the numbers **2, 7,** and **9.**

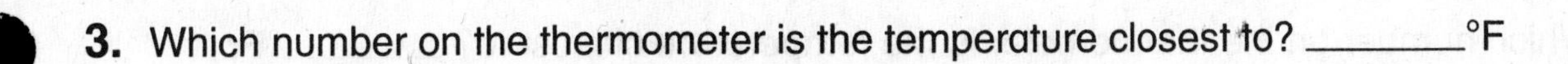

3. Which number on the thermometer is the temperature closest to? ______°F

4. How much money is this? __________

5. Divide each shape in half.

Shade one half of each shape.

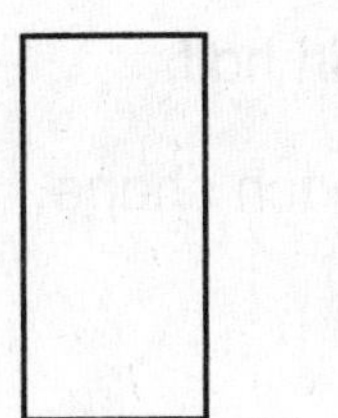

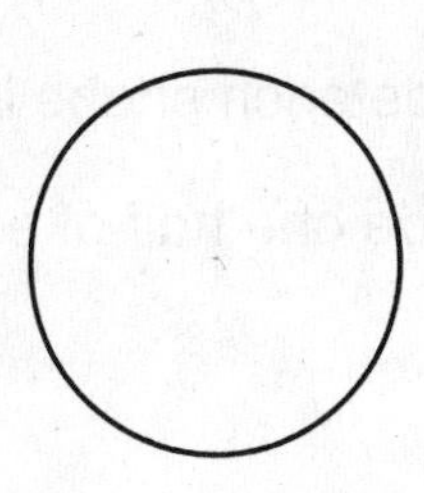

6. Finish the pattern.

△, □, △, ○, △, □, △, ○, △, ______, ______, ______

Name ____________________

Date ____________________

Homework 29B

***Saxon** Math 2 (for use with **Lesson 29**)*

1. Marsha and Susan went to the movies with three friends. How many children went to the movies together? Draw a picture and write a number sentence for this story. Write the answer with a label.

Number sentence ____________________

Answer ____________________

2. Write an addition and subtraction fact family using the numbers **1, 3,** and **4.**

________ ________ ________ ________

3. Which number on the thermometer is the temperature closest to? ______°F

°F 110 100 90 80 70 60 50 40 30 20 10 0 –10 –20

4. How much money is this? __________

5. Divide each shape in half.

Shade one half of each shape.

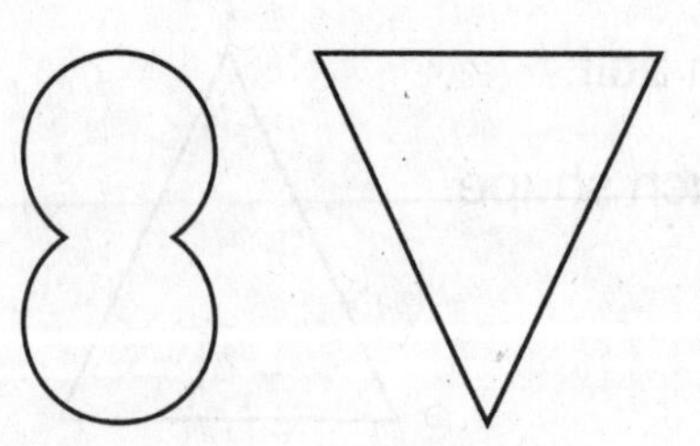

6. What time (to the nearest half hour) do you usually wake up on a school day?

:

Name ____________________ Score ______

Fact Assessment 5

Saxon *Math 2 (for use with* ***Lesson 30-1****)*

Set 5: Doubles + 1

$3 + 4$	$7 + 6$	$4 + 5$	$6 + 5$	$8 + 9$
$8 + 7$	$5 + 6$	$9 + 8$	$4 + 3$	$6 + 7$
$5 + 4$	$7 + 8$	$3 + 4$	$7 + 6$	$5 + 4$
$9 + 8$	$4 + 5$	$7 + 8$	$4 + 3$	$5 + 6$
$8 + 9$	$6 + 5$	$6 + 7$	$8 + 7$	$3 + 4$

Name ______________________________

***Saxon** Math 2 (for use with **Lesson 30-1**)*

A.

0	1	2	3	4
+ ☐	+ ☐	+ ☐	+ ☐	+ ☐
8	8	8	8	8

8	7	6	5	4
+ ☐	+ ☐	+ ☐	+ ☐	+ ☐
8	8	8	8	8

B.

0	1	2	3	4
+ ☐	+ ☐	+ ☐	+ ☐	+ ☐
9	9	9	9	9

9	8	7	6	5
+ ☐	+ ☐	+ ☐	+ ☐	+ ☐
9	9	9	9	9

C.

3 + 5 = ________ 3 + 6 = ________

6 + 3 = ________ 5 + 3 = ________

Dear Parent,

Today we practiced the sum of 8 and 9 facts. Your child will be tested next week on the new addition facts listed below.

$$\begin{array}{r} 3 \\ +\ 5 \\ \hline \end{array} \qquad \begin{array}{r} 5 \\ +\ 3 \\ \hline \end{array} \qquad \begin{array}{r} 3 \\ +\ 6 \\ \hline \end{array} \qquad \begin{array}{r} 6 \\ +\ 3 \\ \hline \end{array}$$

During the next four days, your child will practice these facts in class both orally and in writing. The back of each day's fact sheet will contain the fact practice homework.

Tonight's fact practice is the following:

1. Ask your child to point to the problems in order and to say the answers.
2. Point to the problems in random order and ask your child to say the answers.

Keep this paper at home to use for practice.

Name ____________________

Date ____________________

Problem-Solving Worksheet 30A

Saxon Math 2 (for use with **Lesson 30-1**)

Understand	Plan	Solve	Check

Look for a Pattern

Draw a Picture

The beads in Leila's bracelet are in a blue, yellow, red repeating pattern. Show the color of the tenth bead on her bracelet.

What color is the tenth bead on Leila's bracelet? ____________________

Name ____________________

Performance Task Worksheet 30B

***Saxon** Math 2 (for use with **Lesson 30-1**)*

Understand	Plan	Solve	Check

The beads in Kamille's necklace are in a red, blue, blue, yellow repeating pattern. Show the color of the twelfth bead on her necklace.

What color is the twelfth bead in Kamille's necklace? ____________________

Circle the problem-solving strategies you used to solve this problem.

Act It Out ***Use Logical Reasoning***

Draw a Picture ***Look for a Pattern***

Explain how you got your answer: ____________________

Name ______________________________

Date ______________________________

Written Assessment 5

***Saxon** Math 2 (for use with **Lesson 30-2**)*

1. Matthew wrote a five-page story. On the next day he added two pages to his story. How long is his story now? Draw a picture and write a number sentence for this story. Write the answer with a label.

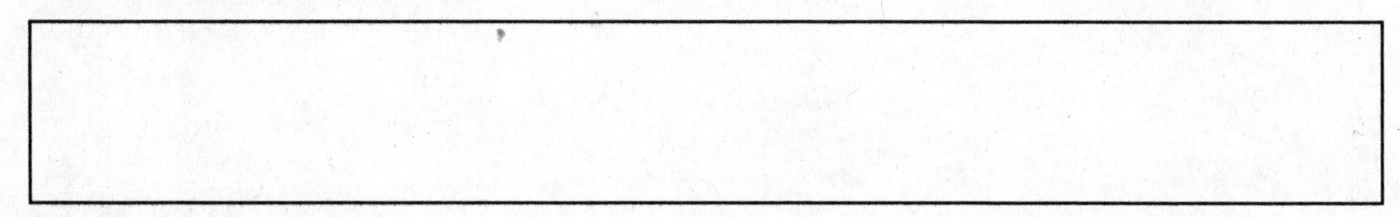

Number sentence ______________________________

Answer ____________________

2. Which number on the thermometer is the temperature closest to? ________°F

°F 110 100 90 80 70 60 50 40 30 20 10 0 −10 −20

3. Divide each shape in half. Shade one half of each shape.

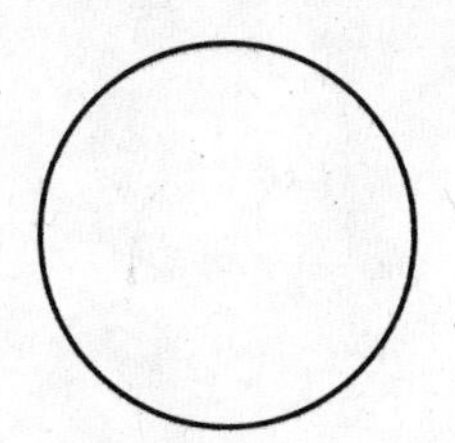
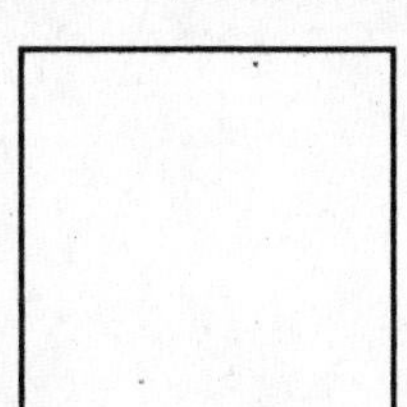

4. Circle the largest number. 17 34 29

5. Match each name with the correct picture.

one half •

one third •

one sixth •

• 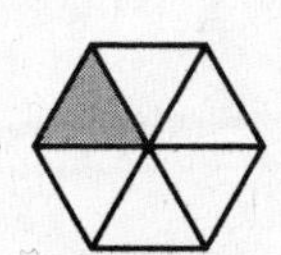

• 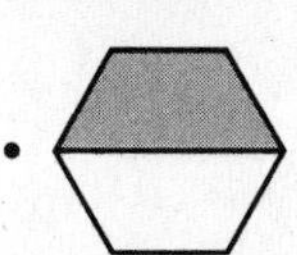

• 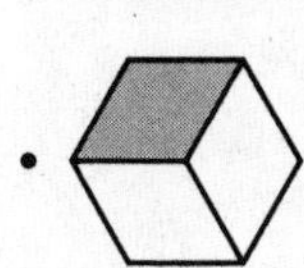

6. Add.

7 + 2 = ______ 3 + 9 = ______ 9 + 6 = ______

2 + 8 = ______ 9 + 2 = ______ 5 + 9 = ______

Name ______________________________

Class Fact Practice 31A

***Saxon** Math 2 (for use with **Lesson 31**)*

Set 6: Sums of 8 and 9; Review Facts

7 + 1	5 + 2	6 + 3	2 + 4	3 + 5
6 + 5	9 + 0	4 + 4	1 + 7	2 + 6
5 + 3	8 + 9	1 + 8	3 + 2	7 + 2
4 + 3	0 + 8	5 + 4	6 + 7	8 + 1
4 + 5	6 + 2	8 + 7	3 + 6	2 + 7

Name ______________________

Fact Homework 31B

***Saxon** Math 2 (for use with **Lesson 31**)*

Set 6: Sums of 8 and 9; Review Facts

1. Read the answers to someone.
2. Write the answers.
3. Ask someone to correct your paper. Corrected by ______________________

2 + 7 =	3 + 6 =	8 + 7 =	6 + 2 =	4 + 5 =
8 + 1 =	6 + 7 =	5 + 4 =	0 + 8 =	4 + 3 =
7 + 2 =	3 + 2 =	1 + 8 =	8 + 9 =	5 + 3 =
2 + 6 =	1 + 7 =	4 + 4 =	9 + 0 =	6 + 5 =
3 + 5 =	2 + 4 =	6 + 3 =	5 + 2 =	7 + 1 =

Name ______________________________

Our Weekday Morning Wake-Up Times

5:00	**5:30**	**6:00**	**6:30**	**7:00**	**7:30**	**8:00**	**8:30**

Name ______________________________

Date ______________________________

Guided Class Practice 31A

***Saxon** Math 2 (for use with **Lesson 31**)*

1. Tim had ten dimes. He gave three to his brother. How many dimes does Tim have now? Draw a picture and write a number sentence for this story. Write the answer with a label.

Number sentence ______________________________

Answer ____________________

2. How much money is this? __________

3. Number the clockface.
Show half past eleven on both clocks.

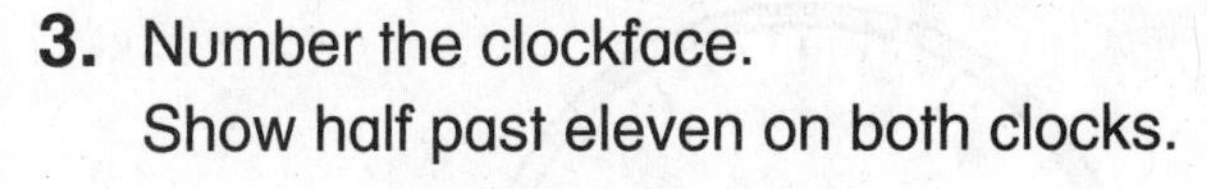

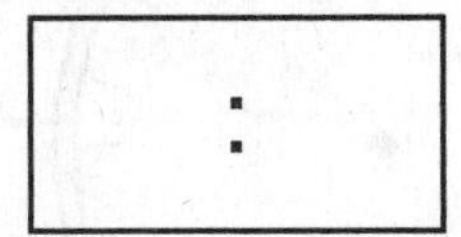

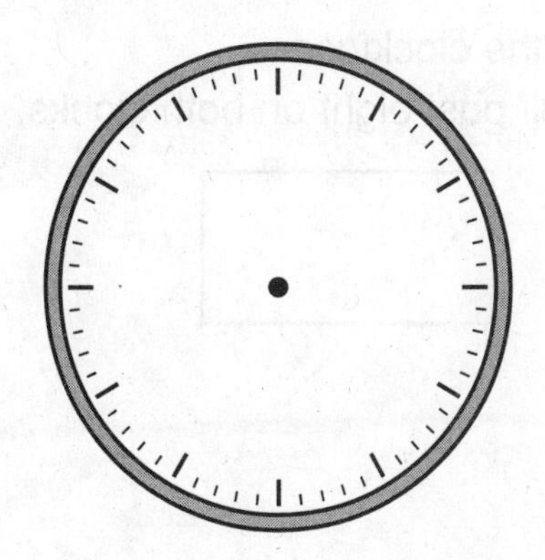

4. The children in Mrs. Abbott's class made this graph.

How many children wake up at half past seven? ________

How many more children wake up at 7:00 than at 6:00? ________

Our Weekday Wake-Up Times

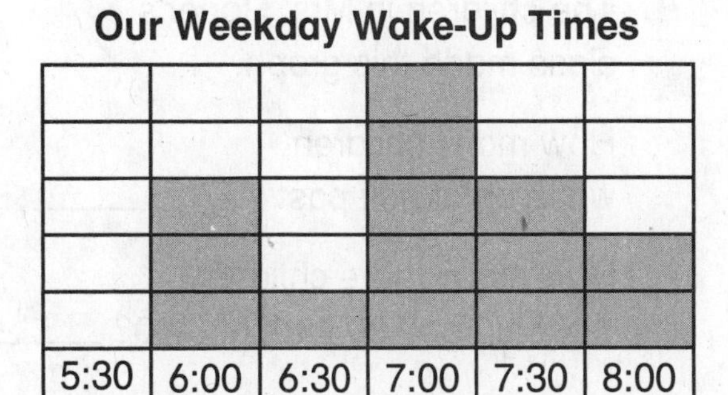

5. Write an addition and subtraction fact family using the numbers **5, 6,** and **11.**

__________ __________ __________ __________

Name ______________________________

Date ______________________________

Homework 31B

***Saxon** Math 2 (for use with **Lesson 31**)*

1. Angelo borrowed six books from the library. He read four books and took them back to the library. How many more books does he have left to read? Draw a picture and write a number sentence for this story. Write the answer with a label.

Number sentence ______________________________

Answer ______________________

2. How much money is this? __________

3. Number the clockface.
Show half past eight on both clocks.

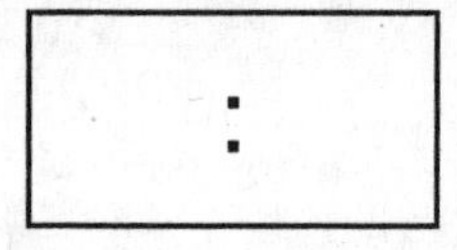

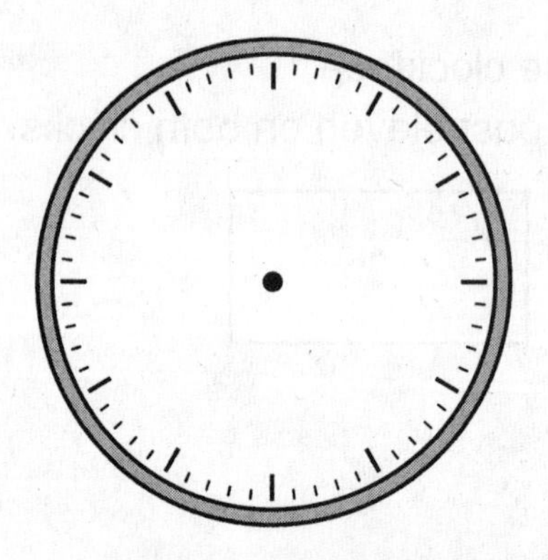

4. The children in Mrs. Roper's class made this graph.

How many children wake up at half past six? ________

How many more children wake up at 7:30 than at 8:00? ________

Our Weekday Wake-Up Times

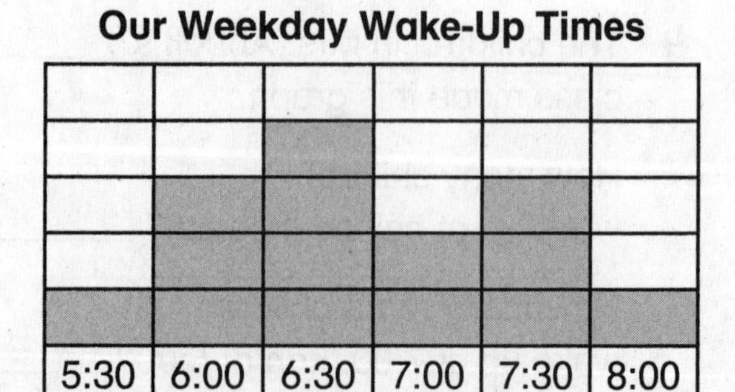

5. Write an addition and subtraction fact family using the numbers **2, 5,** and **7.**

__________ __________ __________ __________

Name ____________________________________

Set 6: Sums of 8 and 9; Review Facts

$\begin{array}{r} 5 \\ +\ 2 \\ \hline \end{array}$	$\begin{array}{r} 6 \\ +\ 3 \\ \hline \end{array}$	$\begin{array}{r} 1 \\ +\ 7 \\ \hline \end{array}$	$\begin{array}{r} 4 \\ +\ 2 \\ \hline \end{array}$	$\begin{array}{r} 3 \\ +\ 5 \\ \hline \end{array}$
$\begin{array}{r} 9 \\ +\ 8 \\ \hline \end{array}$	$\begin{array}{r} 4 \\ +\ 5 \\ \hline \end{array}$	$\begin{array}{r} 2 \\ +\ 3 \\ \hline \end{array}$	$\begin{array}{r} 3 \\ +\ 6 \\ \hline \end{array}$	$\begin{array}{r} 7 \\ +\ 6 \\ \hline \end{array}$
$\begin{array}{r} 7 \\ +\ 2 \\ \hline \end{array}$	$\begin{array}{r} 5 \\ +\ 3 \\ \hline \end{array}$	$\begin{array}{r} 8 \\ +\ 7 \\ \hline \end{array}$	$\begin{array}{r} 2 \\ +\ 6 \\ \hline \end{array}$	$\begin{array}{r} 5 \\ +\ 4 \\ \hline \end{array}$
$\begin{array}{r} 3 \\ +\ 4 \\ \hline \end{array}$	$\begin{array}{r} 0 \\ +\ 9 \\ \hline \end{array}$	$\begin{array}{r} 7 \\ +\ 1 \\ \hline \end{array}$	$\begin{array}{r} 5 \\ +\ 6 \\ \hline \end{array}$	$\begin{array}{r} 2 \\ +\ 5 \\ \hline \end{array}$
$\begin{array}{r} 6 \\ +\ 2 \\ \hline \end{array}$	$\begin{array}{r} 1 \\ +\ 8 \\ \hline \end{array}$	$\begin{array}{r} 7 \\ +\ 8 \\ \hline \end{array}$	$\begin{array}{r} 2 \\ +\ 7 \\ \hline \end{array}$	$\begin{array}{r} 4 \\ +\ 4 \\ \hline \end{array}$

Name ______________________

Set 6: Sums of 8 and 9; Review Facts

1. Read the answers to someone.
2. Write the answers.
3. Ask someone to correct your paper. Corrected by ______________________

4 + 4 =	2 + 7 =	7 + 8 =	1 + 8 =	6 + 2 =
2 + 5 =	5 + 6 =	7 + 1 =	0 + 9 =	3 + 4 =
5 + 4 =	2 + 6 =	8 + 7 =	5 + 3 =	7 + 2 =
7 + 6 =	3 + 6 =	2 + 3 =	4 + 5 =	9 + 8 =
3 + 5 =	4 + 2 =	1 + 7 =	6 + 3 =	5 + 2 =

Name ______________________

Date ______________________

Guided Class Practice 32A

***Saxon** Math 2 (for use with **Lesson 32**)*

1. Shawn has 2 dimes and 5 pennies. Steven has 3 dimes and 4 pennies. How many dimes do the two boys have altogether? Draw a picture and write a number sentence for this story. Write the answer with a label.

Number sentence ______________________

Answer ______________________

2. Draw tally marks to show **13.**

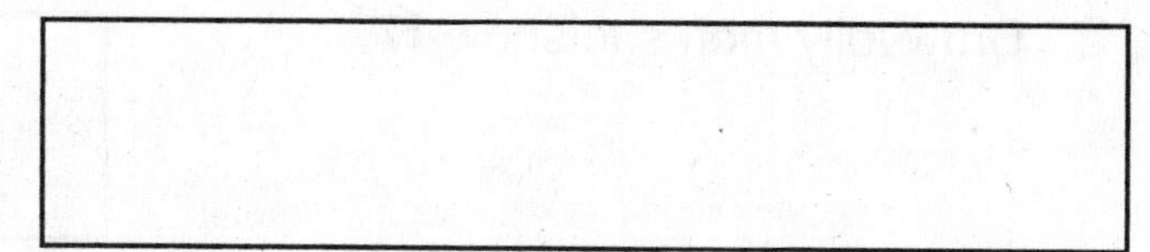

3. Write an addition and subtraction fact family using the numbers **11, 6,** and **5.**

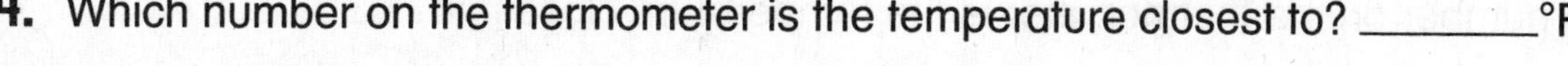

4. Which number on the thermometer is the temperature closest to? ______ °F

°F

110
100
90
80
70
60
50
40
30
20
10
0
−10
−20

5. How much money is this? ______

6. What is the fourth letter? ______

What is the sixth letter? ______

What is the second letter? ______

What is the ninth letter? ______

What is the eleventh letter? ______

T A P S O M N U R Y T E

Name ____________________

Date ____________________

Homework 32B

***Saxon** Math 2 (for use with **Lesson 32**)*

1. Shawn has 2 dimes and 5 pennies. Steven has 3 dimes and 4 pennies. How many pennies do the two boys have altogether? Draw a picture and write a number sentence for this story. Write the answer with a label.

Number sentence ____________________

Answer ____________________

2. Draw tally marks to show **17**.

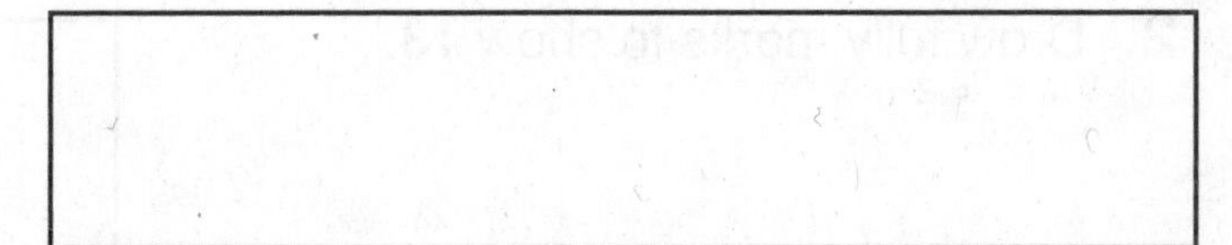

3. Write an addition and subtraction fact family using the numbers **15, 8,** and **7.**

__________ __________ __________ __________

4. Which number on the thermometer is the temperature closest to? ______°F

5. How much money is this? __________

6. What is the sixth letter? ______

What is the fifth letter? ______

What is the seventh letter? ______

What is the twelfth letter? ______

What is the tenth letter? ______

T A P S O M N U R Y T E

Name ______________________

Class Fact Practice 33A

***Saxon** Math 2 (for use with **Lesson 33**)*

Set 6: Sums of 8 and 9

Fill in the missing addends. Draw lines to connect the problems that have the same addends.

A.

1 + ☐ = 8 •	• 3 + ☐ = 8
5 + ☐ = 8 •	• 0 + ☐ = 8
6 + ☐ = 8 •	• 4 + ☐ = 8
8 + ☐ = 8 •	• 7 + ☐ = 8
4 + ☐ = 8 •	• 2 + ☐ = 8

B.

2 + ☐ = 9 •	• 6 + ☐ = 9
5 + ☐ = 9 •	• 1 + ☐ = 9
0 + ☐ = 9 •	• 4 + ☐ = 9
8 + ☐ = 9 •	• 7 + ☐ = 9
3 + ☐ = 9 •	• 9 + ☐ = 9

Name ____________________

Fact Homework 33B

Saxon *Math 2 (for use with* ***Lesson 33****)*

Set 6: Sums of 8 and 9 Corrected by ______

Fill in the missing addends. Draw lines to connect the problems that have the same addends.

A.

☐ + 2 = 8 •	• ☐ + 3 = 8
☐ + 7 = 8 •	• ☐ + 4 = 8
☐ + 5 = 8 •	• ☐ + 1 = 8
☐ + 0 = 8 •	• ☐ + 6 = 8
☐ + 4 = 8 •	• ☐ + 8 = 8

B.

☐ + 2 = 9 •	• ☐ + 8 = 9
☐ + 6 = 9 •	• ☐ + 0 = 9
☐ + 1 = 9 •	• ☐ + 5 = 9
☐ + 4 = 9 •	• ☐ + 7 = 9
☐ + 9 = 9 •	• ☐ + 3 = 9

Name ______________________

Saxon Math 2 (for use with Lesson 33)

Cut along the dotted lines.

31	32	33	34
41	42	43	44
51	52	53	54

1	2	3	4
11	12	13	14
21	22	23	24

25	26	27
35	36	37
45	46	47

8	9	10
18	19	20
28	29	30
38	39	40
48	49	50

61	62	63	64
71	72	73	74
81	82	83	84
91	92	93	94

5	6	7
15	16	17

55	56
65	66
75	76
85	86
95	96

77	78	79	80
87	88	89	90
97	98	99	100

57	58	59	60
67	68	69	70

Name ______________________

Date ______________________

Guided Class Practice 33A

***Saxon** Math 2 (for use with **Lesson 33**)*

1. Scott had 12 balloons. Two popped. How many balloons does he have now? Draw a picture and write a number sentence for this story. Write the answer with a label.

Number sentence ______________________

Answer ______________________

2. Trace the horizontal line segment with a red crayon.
Trace the vertical line segment with a blue crayon.
Trace the oblique line segment with a yellow crayon.

3. Divide each shape in half.
Shade one half of each shape.

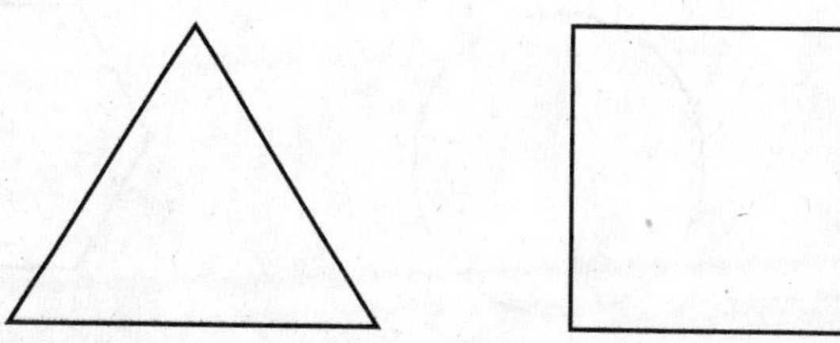

4. Use the Weekday Wake-Up Times class graph to answer these questions.

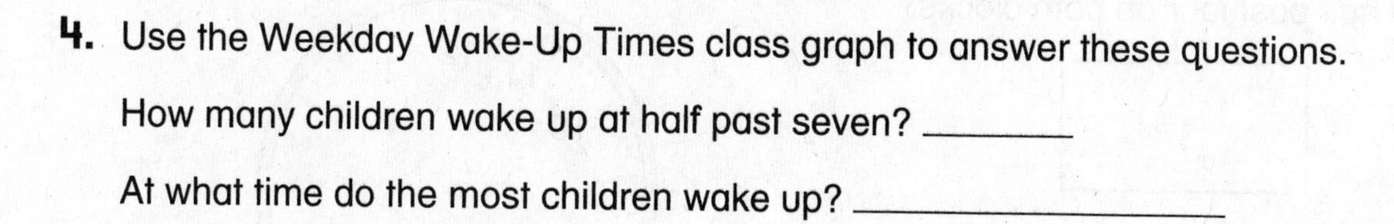

How many children wake up at half past seven? ________

At what time do the most children wake up? ______________

5. How many tally marks are shown? ________

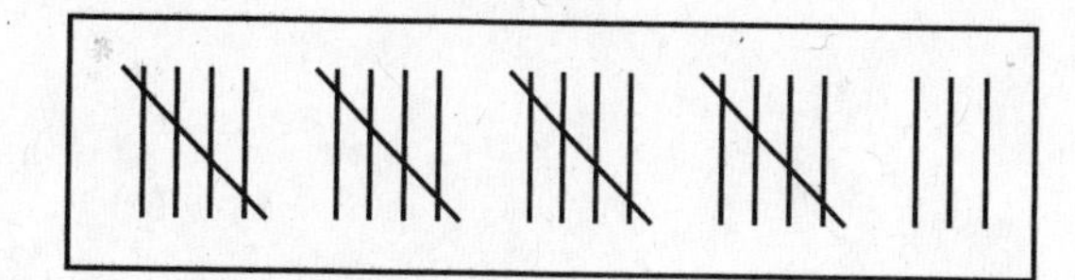

Name ____________________

Date ____________________

Saxon Math 2 (for use with **Lesson 33**)

1. Nicole had seven markers. She gave two markers to her friend Jessica. How many markers does Nicole have now? Draw a picture and write a number sentence for this story. Write the answer with a label.

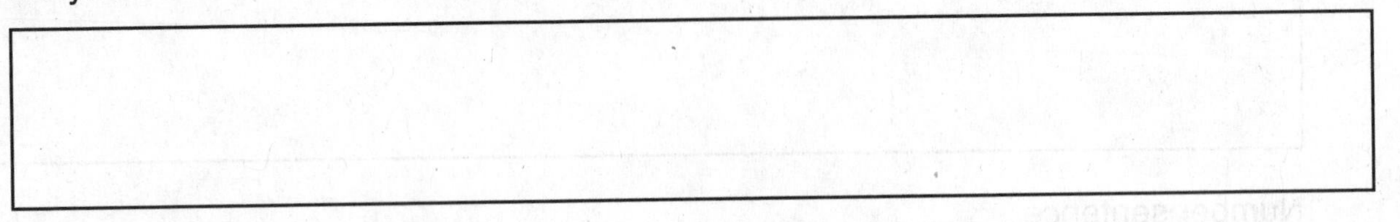

Number sentence ____________________

Answer ____________________

2. Trace the horizontal line segment with a red crayon.
Trace the vertical line segment with a blue crayon.
Trace the oblique line segment with a yellow crayon.

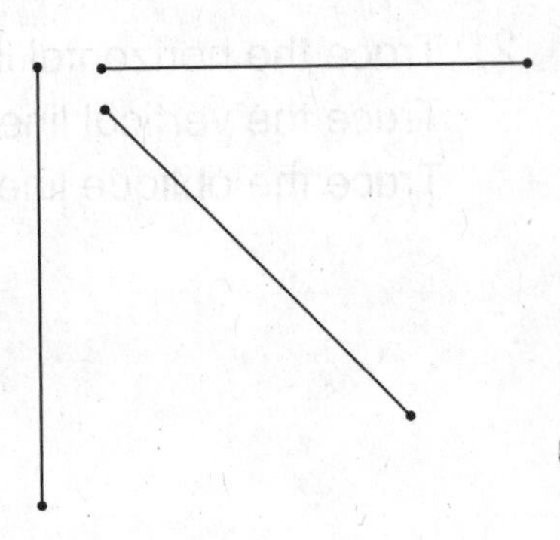

3. Divide each shape in half.
Shade one half of each shape.

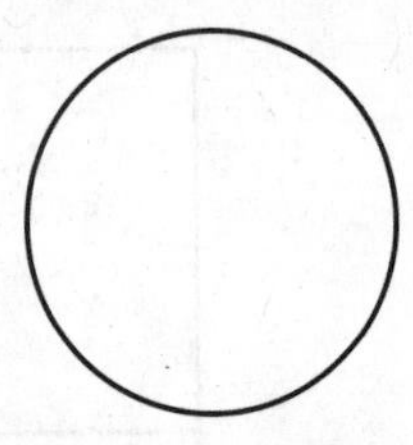
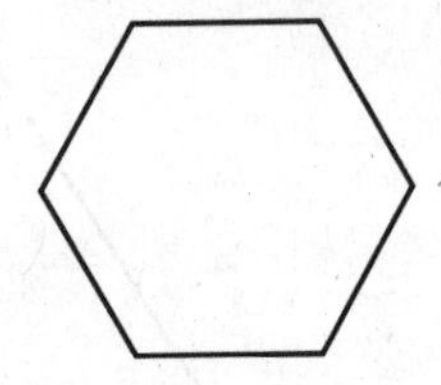
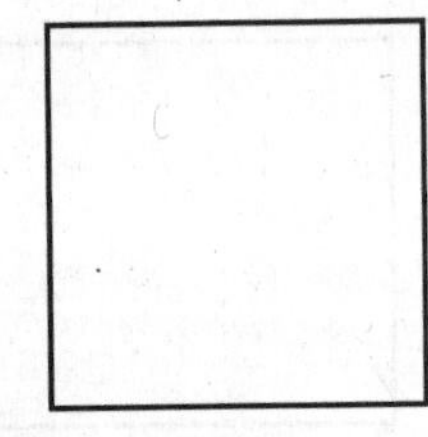

4. Show half past four on both clocks.

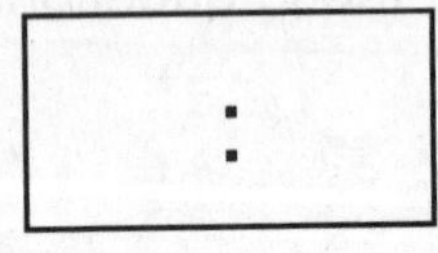

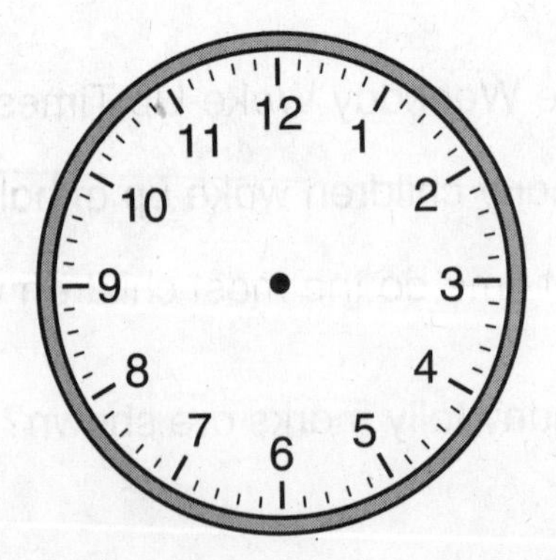

5. How many tally marks are shown? ________

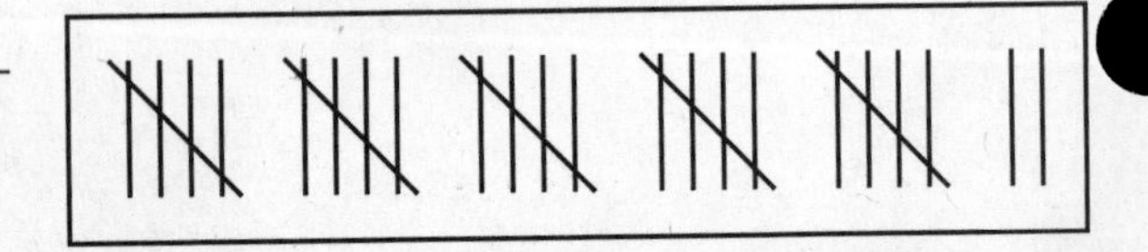

Name ______________________________

Class Fact Practice **34A**

***Saxon** Math 2 (for use with **Lesson 34**)*

Set 6: Sums of 8 and 9; Review Facts

1 + 8	8 + 9	5 + 3	7 + 2	3 + 2
6 + 3	5 + 2	7 + 1	3 + 5	2 + 4
4 + 4	9 + 0	6 + 5	1 + 7	2 + 6
5 + 4	0 + 8	4 + 3	8 + 1	6 + 7
2 + 7	6 + 2	4 + 5	8 + 7	3 + 6

Name ____________________

Fact Homework 34B

***Saxon** Math 2 (for use with **Lesson 34**)*

Set 6: Sums of 8 and 9; Review Facts

1. Read the answers to someone.
2. Write the answers.
3. Ask someone to correct your paper. Corrected by ____________________

8 + 1	4 + 3	0 + 8	5 + 4	6 + 2
2 + 7	6 + 7	4 + 5	8 + 7	3 + 6
1 + 7	9 + 0	2 + 6	6 + 5	8 + 8
2 + 4	3 + 5	5 + 2	7 + 1	6 + 3
3 + 2	7 + 2	5 + 3	8 + 9	1 + 8

Name ______________________________

Name ______________________

Date ______________________

Guided Class Practice 34A

***Saxon** Math 2 (for use with **Lesson 34**)*

1. Each weekday Paul eats a bowl of cereal for breakfast. How many bowls of cereal will Paul eat in one week?

Answer ______________________

2. How many children are in your classroom? ________
Draw tally marks to show this number.

3. Divide the circle into halves.

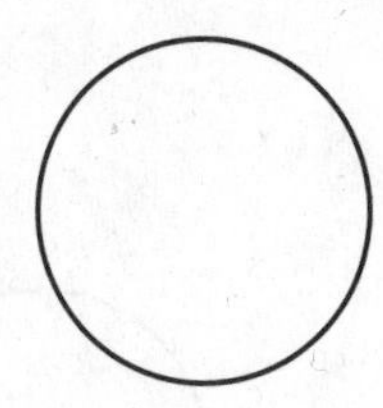

Divide the circle into fourths.

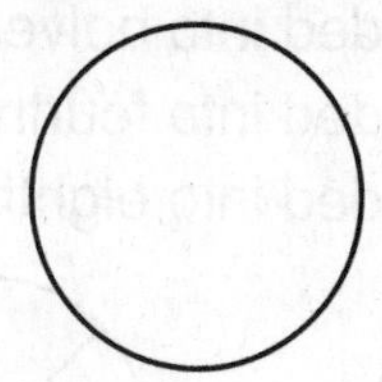

4. It's half past five now.
What time was it one hour ago?
Show this time on both clocks.

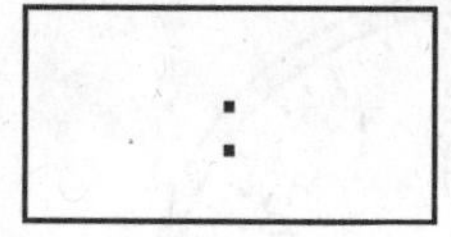

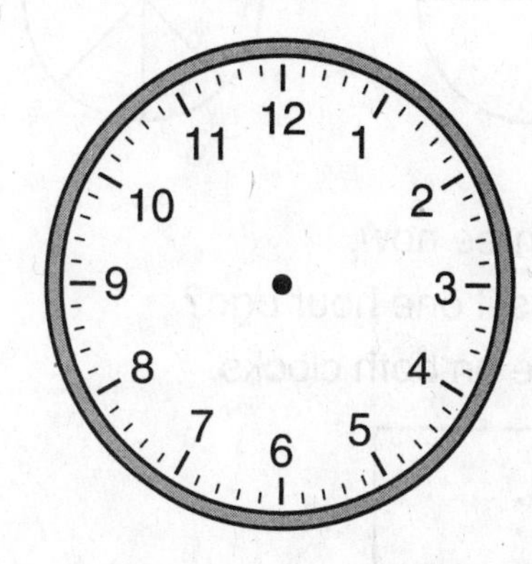

5. Circle the even numbers on the clockface.

6. Write an addition and subtraction fact family using the numbers **3, 8,** and **5.**

__________ __________ __________ __________

7. How much money is 2 dimes? __________

How much money is 30 pennies? __________

Circle the one that is worth more.

Name ______________________________

Date ______________________________

1. Each day of the week Debbie eats a muffin for breakfast. How many muffins will Debbie eat in one week?

Answer ______________________

2. How many people are in your family? ________
Draw tally marks to show this number.

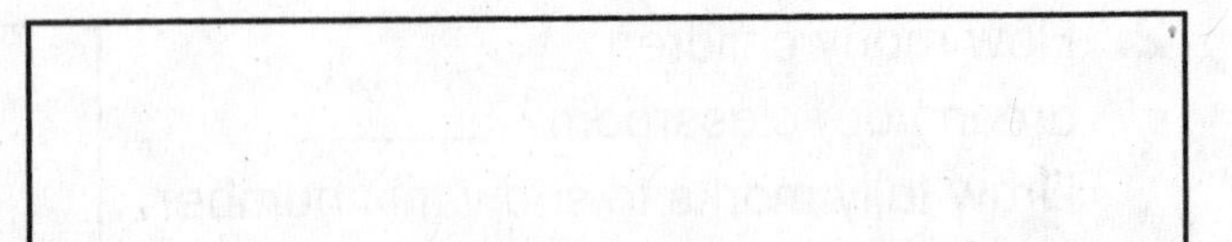

3. Color the whole circle yellow.
Color the circle divided into halves blue.
Color the circle divided into fourths red.
Color the circle divided into eighths green.

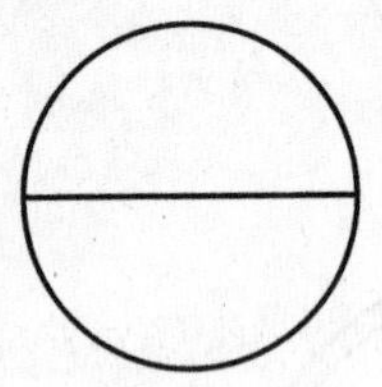
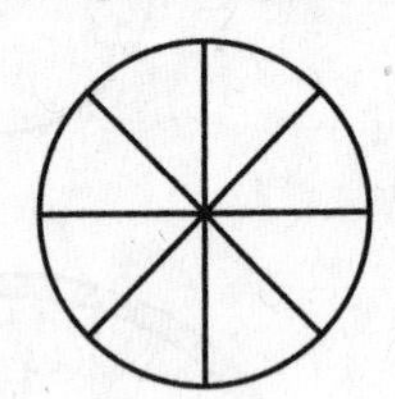
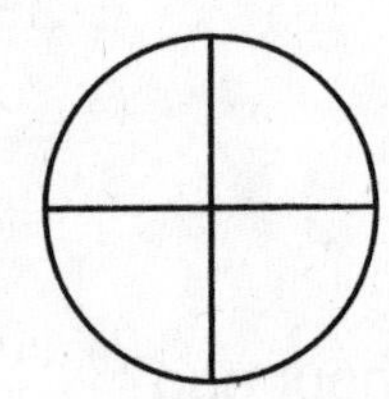
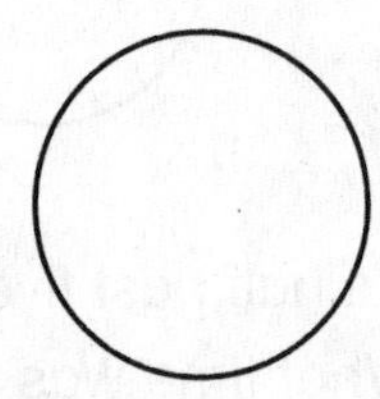

4. It's half past three now.
What time was it one hour ago?
Show this time on both clocks.

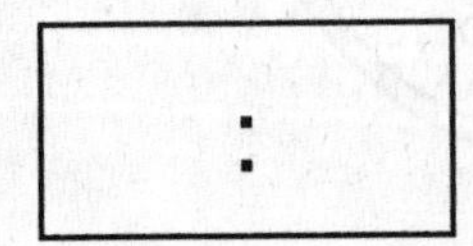

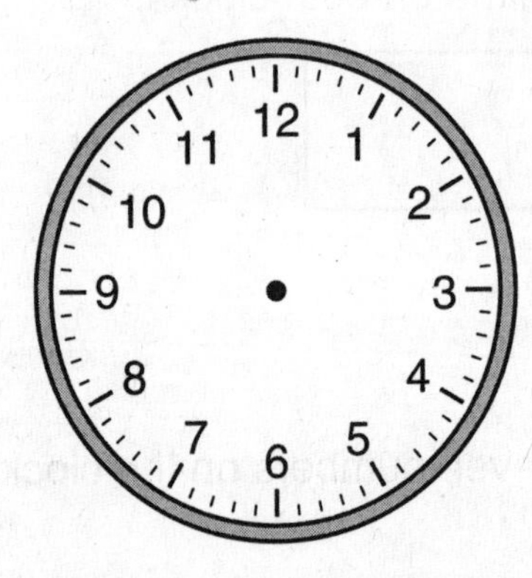

5. Circle the odd numbers on the clockface.

6. Write an addition and subtraction fact family using the numbers **3, 9,** and **6.**

____________ ____________ ____________ ____________

7. How much money is 6 dimes? ____________

How much money is 42 pennies? ____________

Circle the one that is worth more.

Name ______________________ Score ________

Fact Assessment 6

***Saxon** Math 2 (for use with **Lesson 35-1**)*

Set 6: Sums of 8 and 9; Review Facts

$2 + 4$	$3 + 5$	$7 + 1$	$5 + 2$	$6 + 3$
$3 + 2$	$7 + 2$	$5 + 3$	$8 + 9$	$1 + 8$
$2 + 6$	$1 + 7$	$6 + 5$	$9 + 0$	$4 + 4$
$6 + 7$	$8 + 1$	$4 + 3$	$0 + 8$	$5 + 4$
$3 + 6$	$8 + 7$	$4 + 5$	$6 + 2$	$2 + 7$

Name ______________________________

Lesson Worksheet 35-1A

Saxon Math 2 (for use with Lesson 35-1)

3 + □ = 10	8 + □ = 10	□ + 4 = 10	5 + □ = 10	□ + 1 = 10
2 + □ = 10	6 + □ = 10	□ + 5 = 10	□ + 3 = 10	9 + □ = 10
7 + □ = 10	□ + 1 = 10	5 + □ = 10	4 + □ = 10	□ + 8 = 10
9 + □ = 10	□ + 2 = 10	6 + □ = 10	□ + 5 = 10	□ + 7 = 10
1 + □ = 10	□ + 6 = 10	3 + □ = 10	8 + □ = 10	□ + 5 = 10

Dear Parent,

Today we practiced the sum of 10 facts. Your child will be tested next week on the new addition facts listed below.

$3 + 7$	$7 + 3$	$4 + 6$	$6 + 4$

During the next four days, your child will practice these facts in class both orally and in writing. The back of each day's fact sheet will contain the fact practice homework.

Keep this paper at home to use for practice.

Name ______________________________

Date ______________________________

Guided Class Practice 35A

***Saxon** Math 2 (for use with **Lesson 35-1**)*

1. What is the fifth day of the week? ____________

What day of the week was yesterday? ____________

What is the eleventh month of the year? ____________

2. Use a red crayon to trace the horizontal line segments in these letters.
Use a blue crayon to trace the vertical line segments in these letters.
Use a yellow crayon to trace the oblique line segments in these letters.

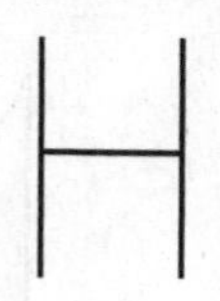

3. Draw a tally mark for each letter of the alphabet.

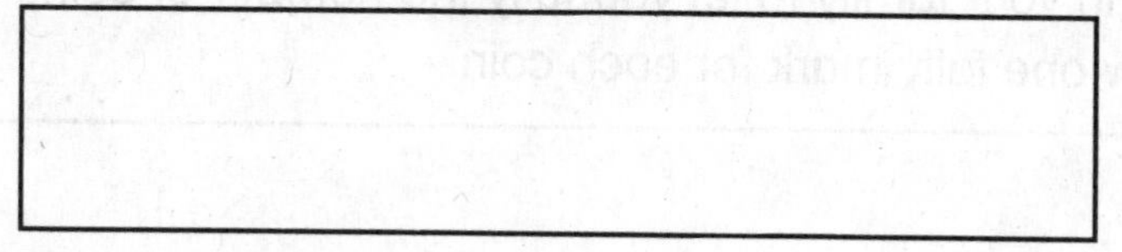

How many tally marks did you draw? ________

4. Fill in the missing addends for the sums of 10.

□ + 2 = 10	1 + □ = 10	□ + 4 = 10
□ + 7 = 10	5 + □ = 10	0 + □ = 10

5. How much money is 6 dimes? ____________

How much money is 54 pennies? ____________

Circle the one that is worth more.

Name ____________________

Date ____________________

Homework 35B

***Saxon** Math 2 (for use with **Lesson 35-1**)*

1. What is the second day of the week? ______________

What day of the week will it be tomorrow? ______________

What is the tenth month of the year? ______________

2. Use a red crayon to trace the horizontal line segments in these letters.
Use a blue crayon to trace the vertical line segments in these letters.
Use a yellow crayon to trace the oblique line segments in these letters.

3. Ask someone in your family to let you tally the number of coins in his or her pocket or wallet. Draw one tally mark for each coin.

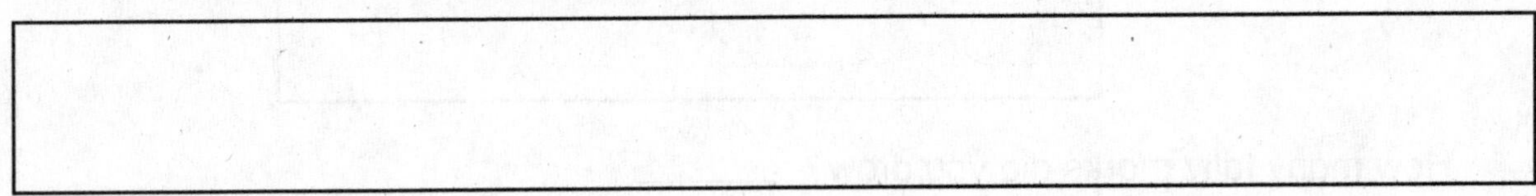

How many tally marks did you draw? ______

4. Fill in the missing addends for the sums of 10.

☐ + 3 = 10	10 + ☐ = 10	☐ + 6 = 10
☐ + 5 = 10	8 + ☐ = 10	9 + ☐ = 10

5. How much money is 7 dimes? ______________

How much money is 39 pennies? ______________

Circle the one that is worth more.

Name ______________________________

Date ______________________________

1. Kathy had eight pencils. Her sister gave her two more. How many pencils does Kathy have now? Draw a picture and write a number sentence for this story. Write the answer with a label.

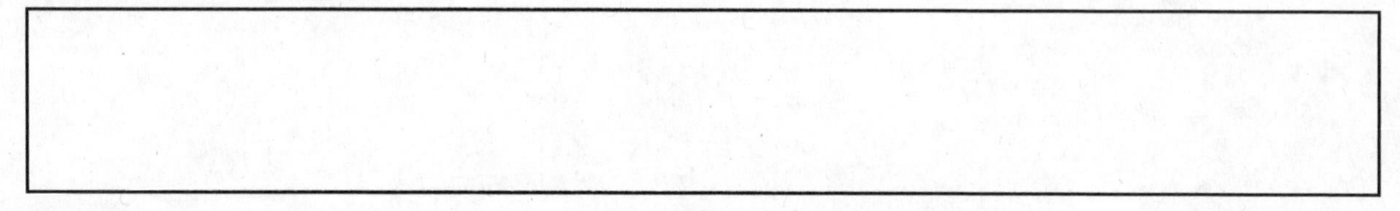

Number sentence ______________________________

Answer ____________________

2. Show two different ways to divide a square in half. Color one half of each square.

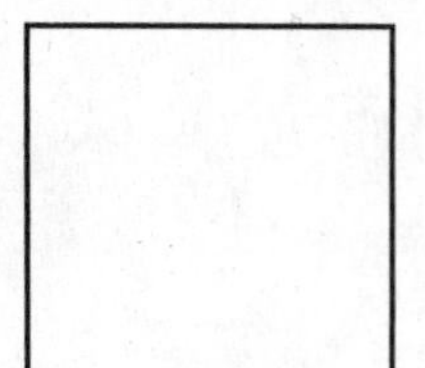
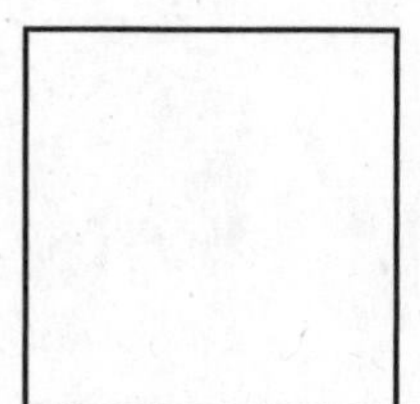

3. How much money is this? __________

4. What is the second day of the week? ____________________

What is the fifth month of the year? ____________________

What is the last month of the year? ____________________

What is the sixth day of the week? ____________________

5. Circle the odd numbers. 4 7 6 3 11 8 12 19

6. Find the sums.

$0 + 4 =$ ____ $4 + 2 =$ ____ $7 + 9 =$ ____ $8 + 8 =$ ____

$4 + 5 =$ ____ $8 + 1 =$ ____ $9 + 4 =$ ____ $7 + 8 =$ ____

Name ______________________________

***Saxon** Math 2 (for use with **Lesson 35-2**)*

Object Weighed	Estimate	Actual

________ ________ ________ ________ ________

lightest heaviest

Name ________________________________

Class Fact Practice 36A

Saxon *Math 2 (for use with* ***Lesson 36****)*

Set 7: Sums of 10; Review Facts

8 + 2	5 + 3	4 + 6	7 + 8	2 + 3
6 + 3	1 + 9	5 + 4	3 + 7	7 + 6
9 + 7	7 + 2	5 + 5	4 + 9	7 + 3
3 + 4	6 + 9	5 + 6	2 + 8	3 + 5
9 + 5	6 + 4	9 + 3	3 + 6	9 + 1

Name ______________________________

Saxon Math 2 (for use with **Lesson 36**)

Set 7: Sums of 10; Review Facts

1. Read the answers to someone.
2. Write the answers.
3. Ask someone to correct your paper. Corrected by ______________________

3 + 5	2 + 8	5 + 6	6 + 9	3 + 4
7 + 3	4 + 9	5 + 5	7 + 2	9 + 7
7 + 6	3 + 7	5 + 4	1 + 9	6 + 3
2 + 3	7 + 8	4 + 6	5 + 3	8 + 2
9 + 1	3 + 6	9 + 3	6 + 4	9 + 5

Name ______________________

Date ______________________

Guided Class Practice 36A

***Saxon** Math 2 (for use with **Lesson 36**)*

1. Six children got on Bus A at the first stop. Five more children got on Bus A at the second stop. How many children are on Bus A now? Draw a picture and write a number sentence for this story. Write the answer with a label.

Number sentence ______________________

Answer ______________________

2. Use the classroom graphs to answer these questions.

How many children wake up at 6:30? ________

How many children's birthdays are in July and August? ________

3. Fill in the missing numbers on this piece of a hundred number chart.

61			64
71			
81		83	

4. Write three addition facts that have sums that are even numbers.

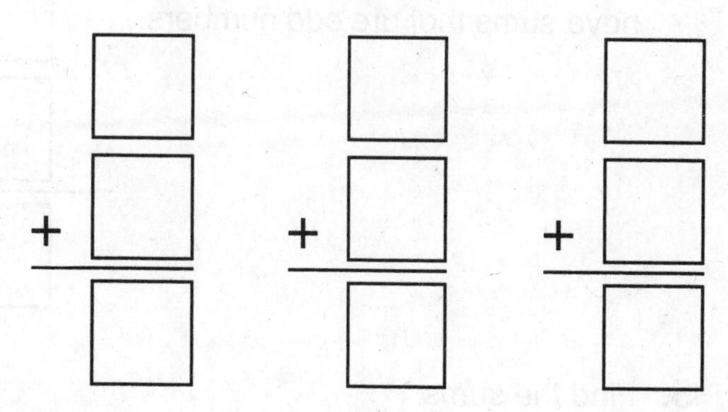

5. Find the sums.

60 + 10 = ________ 40 + 10 = ________ 10 + 80 = ________

6. Draw a horizontal line. Draw a vertical line. Draw an oblique line.

Name ______________________

Date ______________________

1. Seven children got on Bus B at the first stop. Six more children got on Bus B at the second stop. How many children are on Bus B now? Draw a picture and write a number sentence for this story. Write the answer with a label.

Number sentence ______________________

Answer ______________________

2. Which number on the thermometer is the temperature closest to? ______ °F

3. Fill in the missing numbers on this piece of a hundred number chart.

77			
87		89	
97			100

4. Write three addition facts that have sums that are odd numbers.

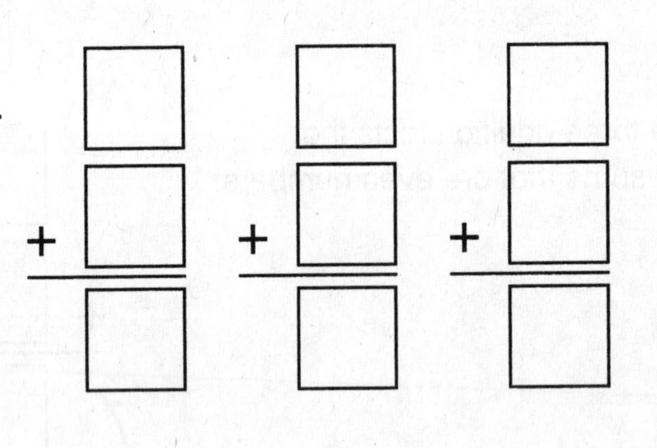

5. Find the sums.

30 + 10 = ______ 10 + 70 = ______ 20 + 10 = ______

6. Draw a horizontal line. Draw a vertical line. Draw an oblique line.

Name ______________________________

Class Fact Practice 37A

Saxon Math 2 (for use with Lesson 37)

Set 7: Sums of 10; Review Facts

$9 + 5$	$6 + 4$	$9 + 7$	$5 + 4$	$3 + 7$
$5 + 5$	$7 + 8$	$5 + 3$	$2 + 8$	$4 + 9$
$3 + 4$	$9 + 1$	$3 + 6$	$9 + 3$	$4 + 6$
$2 + 3$	$6 + 9$	$8 + 2$	$7 + 6$	$3 + 5$
$7 + 3$	$5 + 6$	$7 + 2$	$1 + 9$	$6 + 3$

Name ______________________________

Fact Homework 37B

***Saxon** Math 2 (for use with **Lesson 37**)*

Set 7: Sums of 10; Review Facts

1. Read the answers to someone.
2. Write the answers.
3. Ask someone to correct your paper. Corrected by ______________________

3 + 4	1 + 9	5 + 3	9 + 5	6 + 4
3 + 6	7 + 8	3 + 7	5 + 6	3 + 5
9 + 3	2 + 8	4 + 9	7 + 6	9 + 1
2 + 3	7 + 3	9 + 7	5 + 5	7 + 2
4 + 6	5 + 4	6 + 9	8 + 2	6 + 3

Name ______________________

Date ______________________

Guided Class Practice 37A

Saxon Math 2 (for use with Lesson 37)

1. Judith had 10 hair ribbons and 3 barrettes. She gave her sister two hair ribbons. How many hair ribbons does she have now? Draw a picture and write a number sentence for this story. Write the answer with a label.

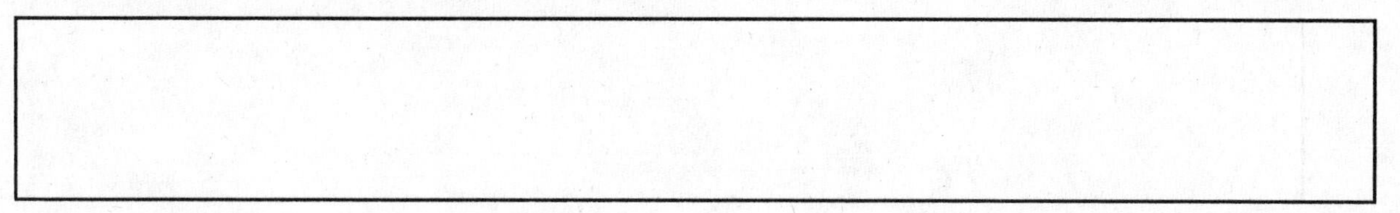

Number sentence ______________________

Answer ______________________

2. How many socks are in the box? ______ socks

Circle pairs of socks.

How many pairs of socks are there? ______ pairs

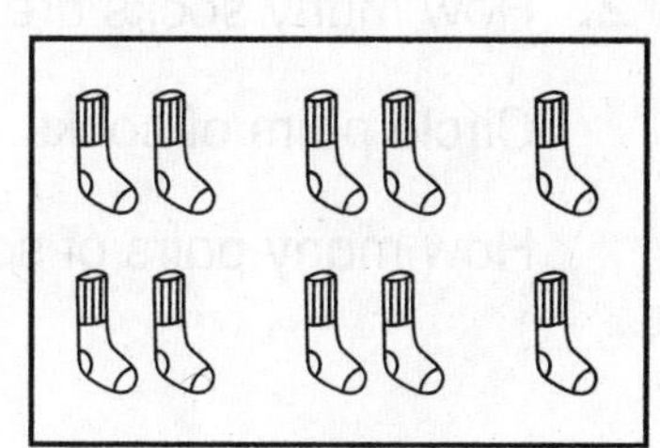

3. Fill in the missing numbers on this piece of a hundred number chart.

23			26		
		35			
43					48

4. Divide these shapes into fourths.

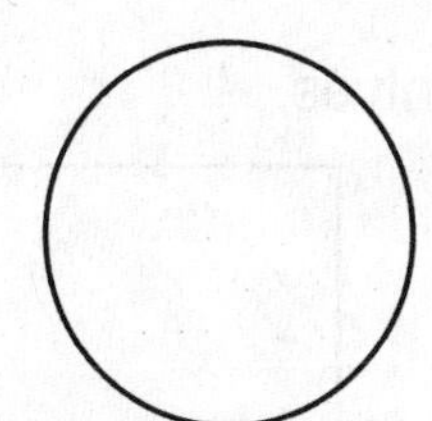

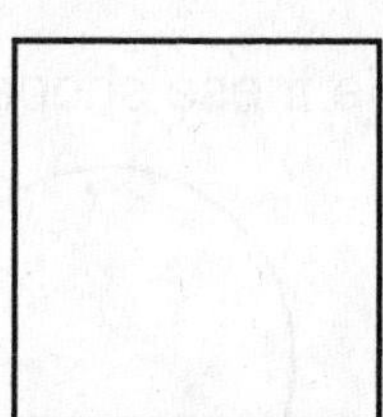

5. Find the sums. $20 + 10 =$ ______ $70 + 10 =$ ______

6. Fill in the missing addends. $\square + 2 = 10$ $6 + \square = 10$

Name ______________________________

Date ______________________________

Homework 37B

***Saxon** Math 2 (for use with **Lesson 37**)*

1. Rebecca's dog ate 15 dog biscuits last week. This week he ate 10 dog biscuits. How many dog biscuits did he eat altogether? Draw a picture and write a number sentence for this story. Write the answer with a label.

Number sentence ______________________________

Answer ______________________________

2. How many socks are in the box? ________ socks

Circle pairs of socks.

How many pairs of socks are there? ________ pairs

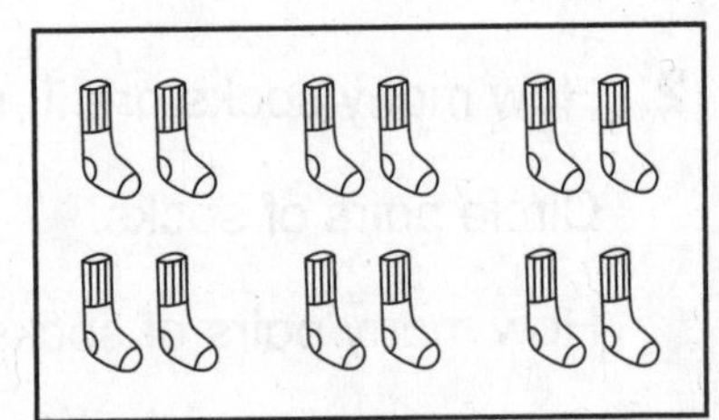

3. Fill in the missing numbers on this piece of a hundred number chart.

42			
			55
62			
		74	

4. Divide these shapes into halves.

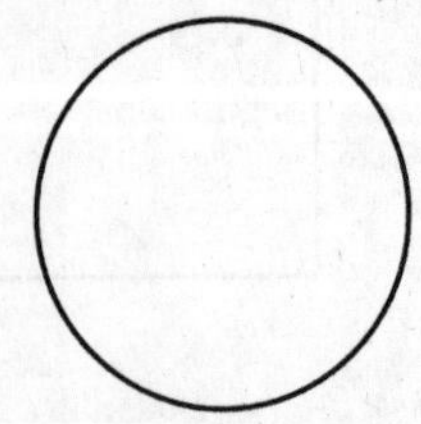

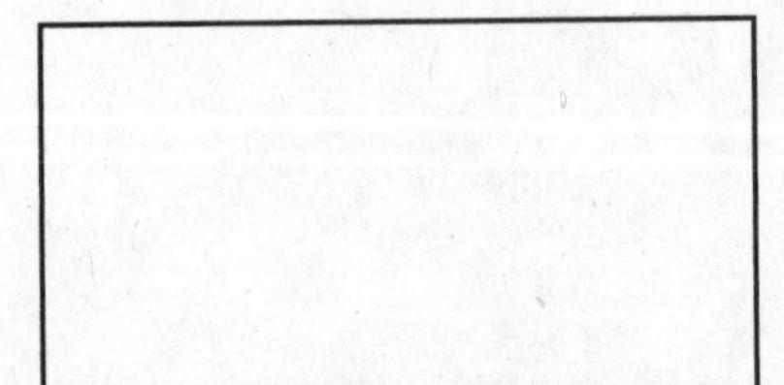

5. Find the sums. 10 + 60 = ________ 30 + 10 = ________

6. Fill in the missing addends. ☐ + 3 = 10 9 + ☐ = 10

Name ______________________________

Set 7: Sums of 10; Review Facts

A.

3 + 7 = ____	7 + 6 = ____	5 + 3 = ____
4 + 6 = ____	9 + 7 = ____	8 + 2 = ____
5 + 6 = ____	9 + 1 = ____	7 + 8 = ____
4 + 9 = ____	7 + 3 = ____	5 + 4 = ____
6 + 9 = ____	6 + 3 = ____	3 + 4 = ____
2 + 8 = ____	9 + 3 = ____	6 + 4 = ____
7 + 2 = ____	1 + 9 = ____	9 + 5 = ____

B.

7 + ☐ = 10	2 + ☐ = 10	4 + ☐ = 10
☐ + 9 = 10	☐ + 3 = 10	☐ + 8 = 10

Name ______________________

Fact Homework 38B

***Saxon** Math 2 (for use with **Lesson** 38)*

Set 7: Sums of 10; Review Facts

A. 1. Read the answers to someone.
2. Write the answers.
3. Ask someone to correct your paper. Corrected by ______________

$6 + 4 =$ ____ $9 + 3 =$ ____ $2 + 8 =$ ____

$9 + 5 =$ ____ $1 + 9 =$ ____ $7 + 2 =$ ____

$3 + 4 =$ ____ $6 + 3 =$ ____ $6 + 9 =$ ____

$5 + 4 =$ ____ $7 + 3 =$ ____ $4 + 9 =$ ____

$7 + 8 =$ ____ $9 + 1 =$ ____ $5 + 6 =$ ____

$8 + 2 =$ ____ $9 + 7 =$ ____ $4 + 6 =$ ____

$5 + 3 =$ ____ $7 + 6 =$ ____ $3 + 7 =$ ____

B. Fill in the missing numbers.

$3 + \square = 10$ $8 + \square = 10$ $6 + \square = 10$

$\square + 1 = 10$ $\square + 7 = 10$ $\square + 2 = 10$

Name ______________________________

Date ______________________________

Guided Class Practice **38A**

***Saxon** Math 2 (for use with **Lesson 38**)*

1. Tom had 3 red markers, 4 green markers, and 5 pencils. He lost 1 pencil. How many pencils does Tom have now? Draw a picture and write a number sentence for this story. Write the answer with a label.

Number sentence ______________________________

Answer ____________________

2. How many shoes are in the box? ______ shoes

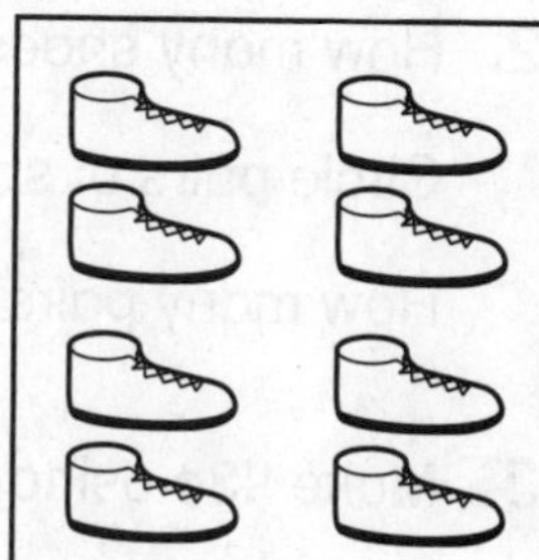

Circle pairs of shoes.

How many pairs of shoes are there? ______ pairs

3. Make 56¢ using the fewest dimes and pennies.

______ dimes ______ pennies

How many tens and ones are in 49?

______ tens ______ ones

4. How much money is this? __________

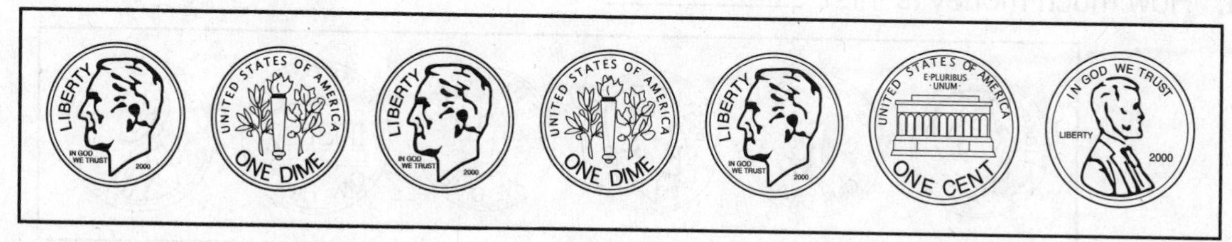

5. Use the Weekday Wake-Up Times class graph to answer these questions.

How many children wake up at 6:00? ______

How many children wake up at 8:00 or later? ______

What is the earliest time that someone in this class wakes up? __________

6. Write an addition and subtraction fact family using the numbers **2, 6,** and **8.**

__________ __________ __________ __________

Name ______________________________

Date ______________________________

Homework 38B

Saxon *Math 2 (for use with* ***Lesson 38****)*

1. Ken's mom gave him 4 chocolate chip cookies, 1 apple, 2 peanut butter cookies, and a drink box for the field trip. How many cookies did she give him? Draw a picture and write a number sentence for this story. Write the answer with a label.

Number sentence ______________________________

Answer ____________________

2. How many shoes are in the box? ______ shoes

Circle pairs of shoes.

How many pairs of shoes are there? ______ pairs

3. Make 43¢ using the fewest dimes and pennies.

______ dimes ______ pennies

How many tens and ones are in 68?

______ tens ______ ones

4. How much money is this? __________

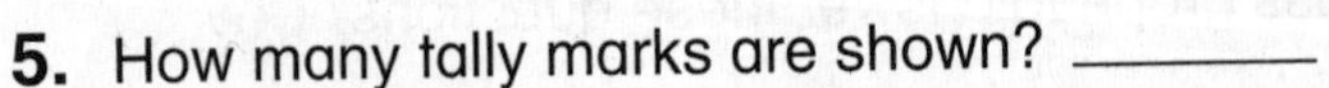

5. How many tally marks are shown? ______

6. Write an addition and subtraction fact family using the numbers **9, 2,** and **11.**

__________ __________ __________ __________

Name ______________________________

Class Fact Practice **39A**

***Saxon** Math 2 (for use with **Lesson 39**)*

Set 7: Sums of 10; Review Facts

9 + 1	7 + 6	4 + 9	2 + 8	9 + 3
3 + 5	5 + 6	3 + 7	7 + 8	3 + 6
6 + 4	9 + 5	5 + 3	1 + 9	3 + 4
6 + 3	8 + 2	6 + 9	5 + 4	4 + 6
7 + 2	5 + 5	9 + 7	7 + 3	2 + 3

Name ______________________

Set 7: Sums of 10; Review Facts

1. Read the answers to someone.
2. Write the answers.
3. Ask someone to correct your paper. Corrected by ______________________

4 + 6	9 + 3	3 + 6	9 + 1	3 + 4
4 + 9	2 + 8	5 + 3	7 + 8	5 + 5
3 + 7	5 + 4	9 + 7	6 + 4	9 + 5
6 + 3	1 + 9	7 + 2	5 + 6	7 + 3
3 + 5	7 + 6	8 + 2	6 + 9	2 + 3

Name ______________________

Date ______________________

Guided Class Practice **39A**

***Saxon** Math 2 (for use with **Lesson 39**)*

1. Anna has 1 ruler, 5 pencils, 2 notebooks, and 4 markers. How many of these things can she use to write with? Draw a picture and write a number sentence for this story. Write the answer with a label.

Number sentence ______________________

Answer ______________________

2. Draw a picture of the favorite apples graph your class made.

Our Favorite Apples

________													
________													
________													

Which apple was chosen by the fewest children? ______________

How many children chose that apple? ________

3. Show **23** using tally marks.

4. Fill in the missing numbers on this piece of a hundred number chart.

		45		
			56	
	64			

5. How much money is 2 pennies and 6 dimes? ________

How much is 5 ones and 3 tens? ________

Name ______________________

Date ______________________

Homework 39B

Saxon *Math 2 (for use with* ***Lesson 39****)*

1. Mike counted three cans of corn, four jars of beans, a package of napkins, and two jars of plant food on the shelf. How many containers of food are safe to eat? Draw a picture and write a number sentence for this story. Write the answer with a label.

Number sentence ______________________

Answer ______________________

2. Use the graph below to answer these questions.

Which apple was the favorite? __________

How many children chose Apple A? __________

How many more children chose Apple A than chose Apple B? __________

Our Favorite Apples

Apple A									
Apple B									
Apple C									

= 1 child

3. Show **17** using tally marks.

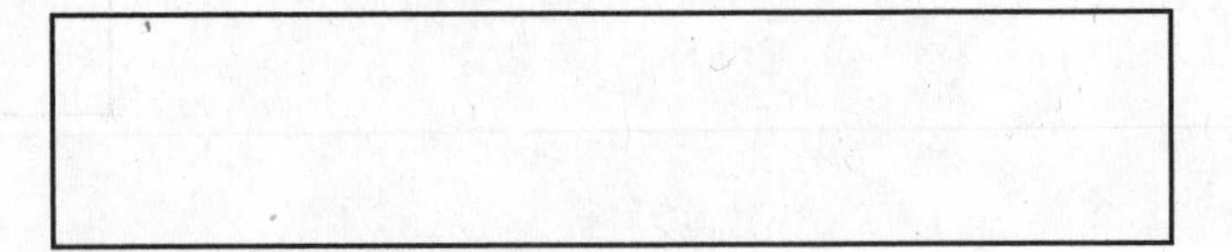

4. Fill in the missing numbers on this piece of a hundred number chart.

	73						
				86			
						98	

5. How much money is 3 pennies and 9 dimes?

____

How much is 6 ones and 2 tens? __________

Name ______________________ Score ______

Fact Assessment 7

***Saxon** Math 2 (for use with **Lesson 40-1**)*

Set 7: Sums of 10; Review Facts

9 + 1	3 + 5	4 + 9	2 + 8	7 + 6
9 + 7	5 + 4	7 + 3	6 + 9	5 + 6
7 + 2	4 + 6	7 + 8	9 + 3	1 + 9
3 + 4	5 + 5	6 + 3	8 + 2	5 + 3
3 + 7	2 + 3	6 + 4	9 + 5	3 + 6

Name ______________________

A.

2 + ☐ = 11	3 + ☐ = 11	4 + ☐ = 11	5 + ☐ = 11	6 + ☐ = 11
9 + ☐ = 11	8 + ☐ = 11	7 + ☐ = 11	6 + ☐ = 11	5 + ☐ = 11

B.

5 + ☐ = 11	7 + ☐ = 11	2 + ☐ = 11	8 + ☐ = 11	4 + ☐ = 11
6 + ☐ = 11	3 + ☐ = 11	9 + ☐ = 11	☐ + 4 = 11	☐ + 8 = 11
☐ + 5 = 11	☐ + 2 = 11	☐ + 7 = 11	☐ + 3 = 11	☐ + 9 = 11

Dear Parent,

Today we practiced the sum of 11 facts. Your child will be tested next week on the new addition facts listed below, along with previously introduced facts.

$$\begin{array}{r} 3 \\ +\ 8 \\ \hline \end{array} \qquad \begin{array}{r} 8 \\ +\ 3 \\ \hline \end{array} \qquad \begin{array}{r} 4 \\ +\ 7 \\ \hline \end{array} \qquad \begin{array}{r} 7 \\ +\ 4 \\ \hline \end{array}$$

During the next four days, your child will practice these facts in class both orally and in writing. The back of each day's fact sheet will contain the fact practice homework.

Keep this paper at home to use for practice.

Name ______________________

Date ______________________

Problem-Solving Worksheet 40A

***Saxon** Math 2 (for use with **Lesson 40-1**)*

Understand	Plan	Solve	Check

Make an Organized List

Act It Out

Henderson has two favorite shirts and two favorite hats. One of his favorite shirts is yellow and the other is red. One of his favorite hats is green and the other is blue. Show all the different ways Henderson can wear his favorite shirts and hats.

How many different ways can Henderson wear his favorite shirts and hats? ________

Name ______________________

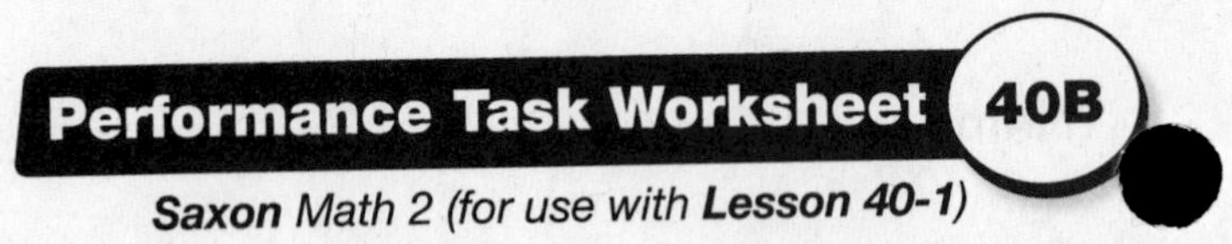

Saxon Math 2 (for use with Lesson 40-1)

Understand	Plan	Solve	Check

William has two favorite shirts and two favorite hats. One of his favorite shirts is blue and the other is red. One of his favorite hats is green and the other is yellow. Show all the different ways he can wear his favorite shirts and hats.

How many different ways can William wear his favorite shirts and hats? ________

Circle the problem-solving strategies you used to solve this problem.

Act It Out ***Use Logical Reasoning***

Draw a Picture ***Look for a Pattern***

Make an Organized List

Explain how you got your answer: ______________________________

____

____

____

Name ______________________________

Date ______________________________

Written Assessment 7

*Saxon Math 2 (for use with **Lesson 40-2**)*

1. Simone helped her brother bake cupcakes. There were 12 cupcakes. They gave two cupcakes to their neighbors. How many cupcakes do they have left? Draw a picture and write a number sentence for this story. Write the answer with a label.

Number sentence ______________________________

Answer ______________________________

2. Number the clockface.
Show half past eight on both clocks.

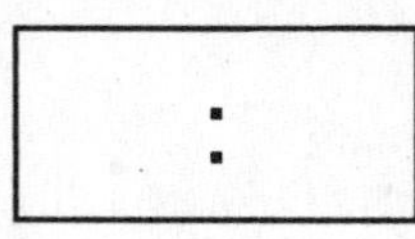

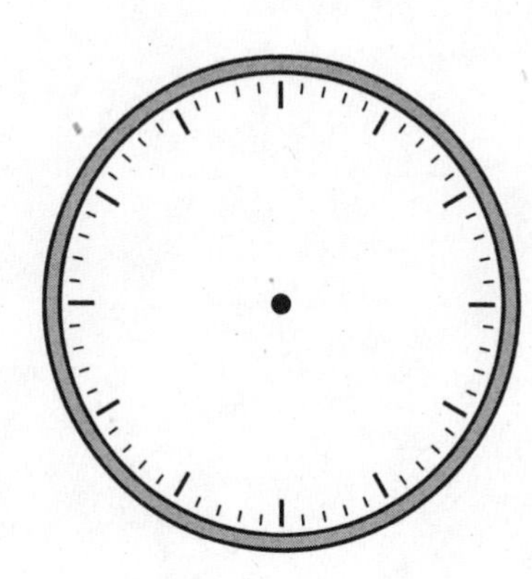

3. Show **32** using tally marks.

4. How much money is this? __________

5. Use the Weekday Wake-Up Times class graph to answer these questions.

How many children wake up at 7:00? ________

At what time do most children wake up? ______________

What is the earliest time that someone in this class wakes up? ______________

6. Write an addition and subtraction fact family using the numbers **4, 5,** and **9.**

__________ __________ __________ __________

Name ______________________________

Lesson Worksheet 40-2

***Saxon** Math 2 (for use with **Lesson 40-2**)*

1\.

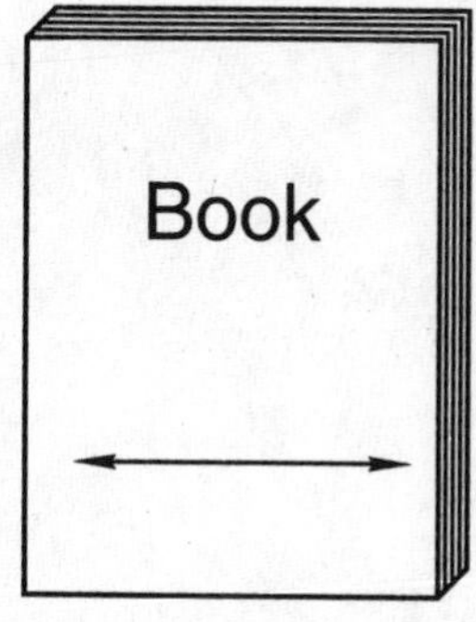

______ inches

2\.

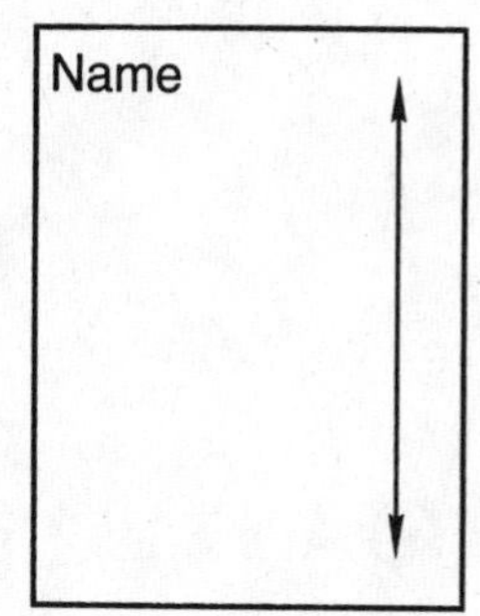

______ inches

3\. Ruler

______ inches

4\. A B

______ inches

5\. C D

______ inches

Name ____________________

Class Fact Practice 41A

***Saxon** Math 2 (for use with **Lesson 41**)*

Set 8: Sums of 11; Review Facts

$8 + 3$	$5 + 9$	$3 + 7$	$9 + 2$	$4 + 3$
$6 + 5$	$9 + 8$	$2 + 9$	$6 + 4$	$4 + 5$
$9 + 6$	$3 + 4$	$5 + 6$	$7 + 3$	$2 + 5$
$7 + 9$	$4 + 6$	$3 + 9$	$6 + 3$	$7 + 4$
$6 + 2$	$4 + 7$	$3 + 5$	$9 + 4$	$3 + 8$

Name ______________________

Fact Homework 41B

Saxon Math 2 (for use with Lesson 41)

Set 8: Sums of 11; Review Facts

1. Read the answers to someone.
2. Write the answers.
3. Ask someone to correct your paper. Corrected by ______________________

$3 + 8$	$9 + 4$	$3 + 5$	$4 + 7$	$6 + 2$
$4 + 5$	$6 + 4$	$2 + 9$	$9 + 8$	$6 + 5$
$7 + 4$	$6 + 3$	$3 + 9$	$4 + 6$	$7 + 9$
$2 + 5$	$7 + 3$	$5 + 6$	$3 + 4$	$9 + 6$
$4 + 3$	$9 + 2$	$3 + 7$	$5 + 9$	$8 + 3$

Name ______________________

Saxon Math 2 (for use with **Lesson 41**)

1.

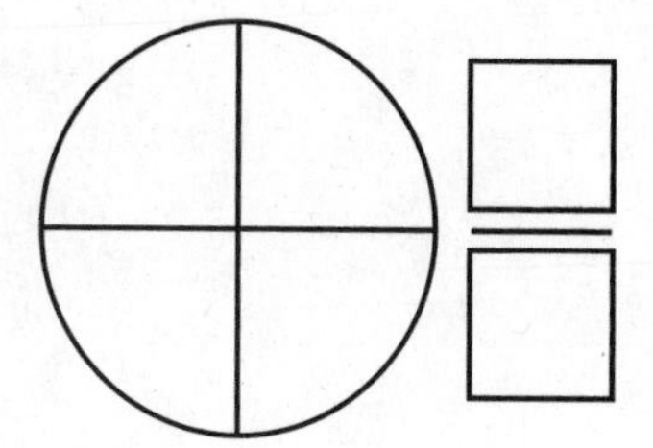

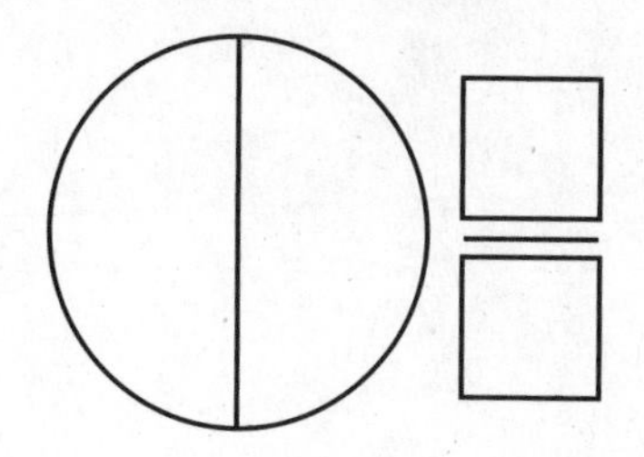

2.

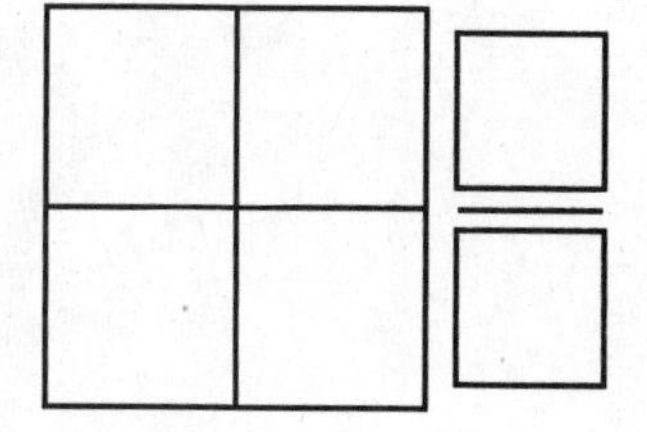

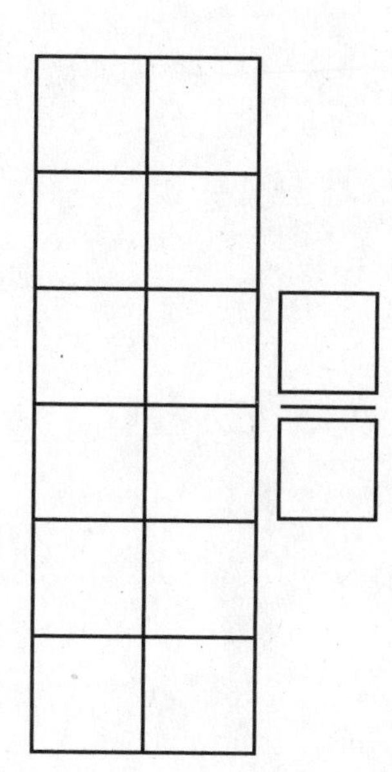

3.

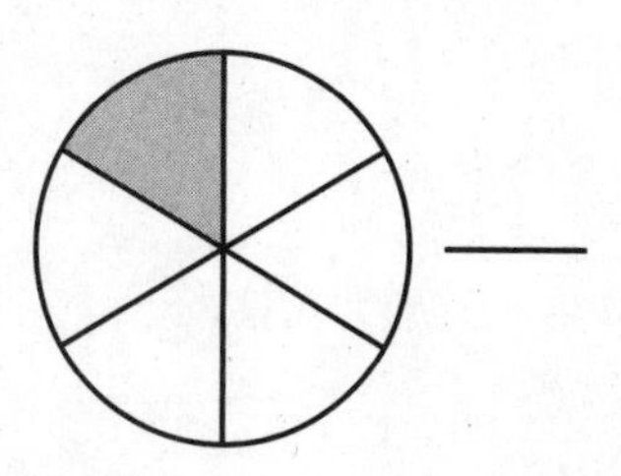

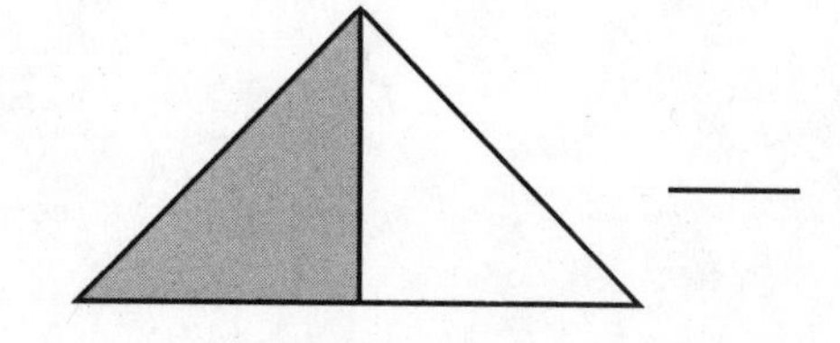

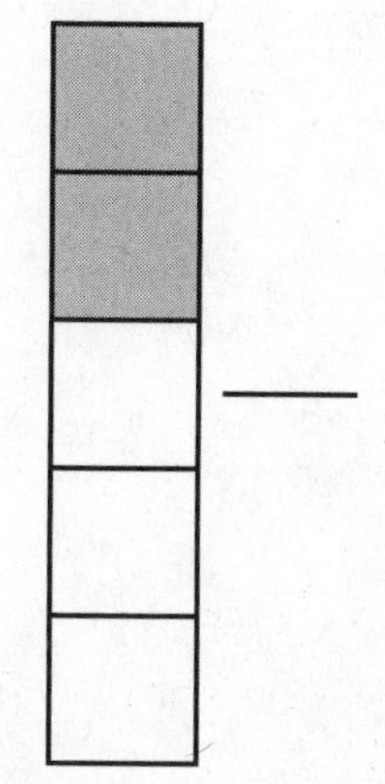

Name ______________________

Date ______________________

Guided Class Practice 41A

***Saxon** Math 2 (for use with **Lesson 41**)*

1. Stephanie and four of her friends were playing outside. Three more friends came to play. How many children are playing now? Draw a picture and write a number sentence for this story. Write the answer with a label.

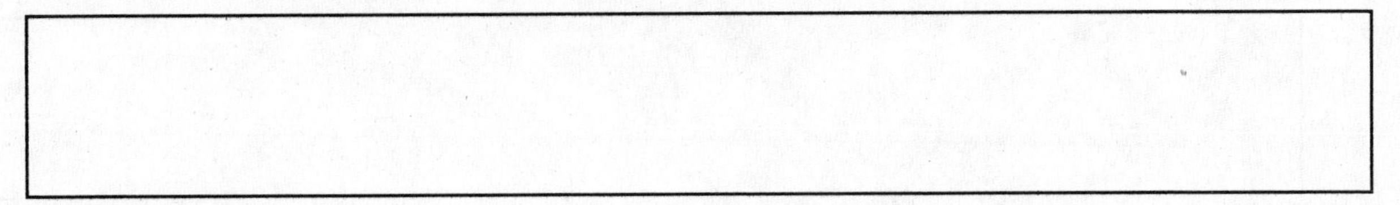

Number sentence ______________________

Answer ______________________

2. Circle pairs of socks.

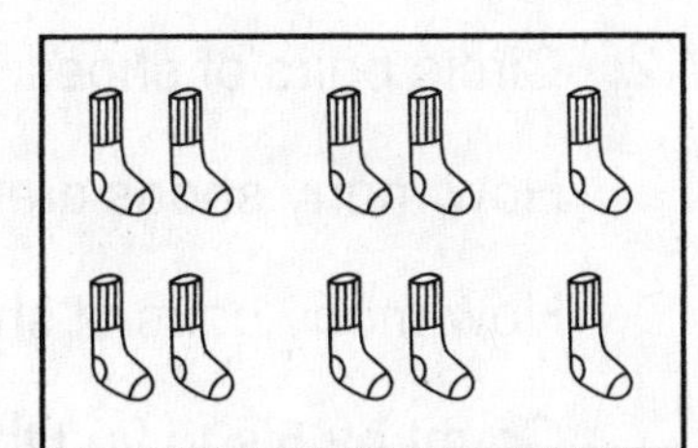

How many socks are in the box? ______ socks

How many pairs of socks are there? ______ pairs

Count by 5's to find the number of toes in all of the socks. ______ toes

3. Shade two fourths. Write the fraction.

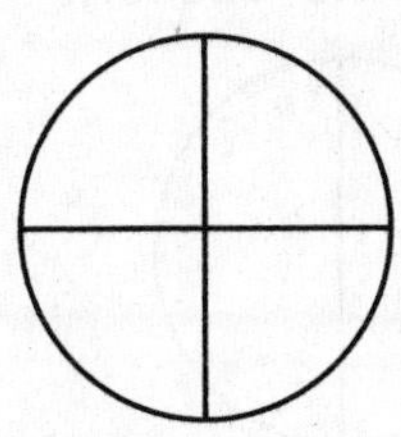

______

Shade four eighths. Write the fraction.

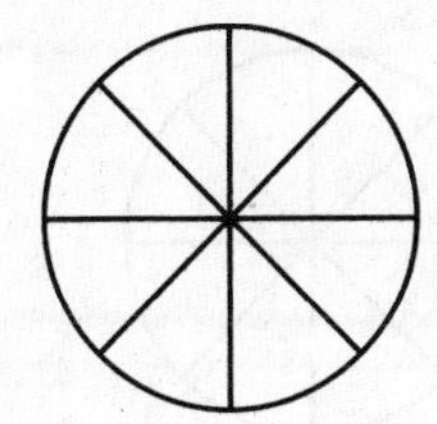

______

Shade one half. Write the fraction.

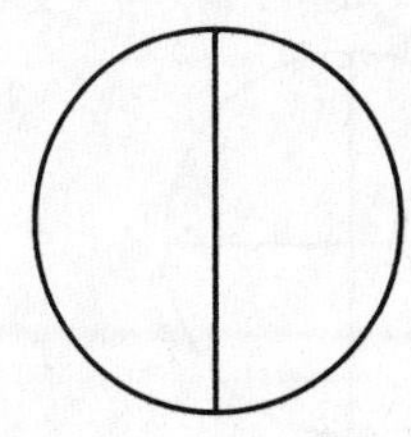

______

4. Make 73¢ using the fewest dimes and pennies.

______ dimes ______ pennies

Make 16 using the fewest tens and ones.

______ ten ______ ones

5. Count by 5's. 5, 10, 15, ______, ______, ______, ______, ______

6. Find the sums.

60 + 10 = ______ 30 + 10 = ______ 10 + 80 = ______

Name ________________________

Date ________________________

Homework 41B

***Saxon** Math 2 (for use with **Lesson 41**)*

1. Colin and three of his friends were making a puzzle. Colin's two sisters came to help. How many children are working on the puzzle now? Draw a picture and write a number sentence for this story. Write the answer with a label.

Number sentence ________________________

Answer ________________________

2. Circle pairs of shoes.

How many shoes are in the box? _______ shoes

How many pairs of shoes are there? _______ pairs

Count by 5's to find the number of toes in all of the shoes. _______ toes

3. Shade three fourths. Write the fraction.

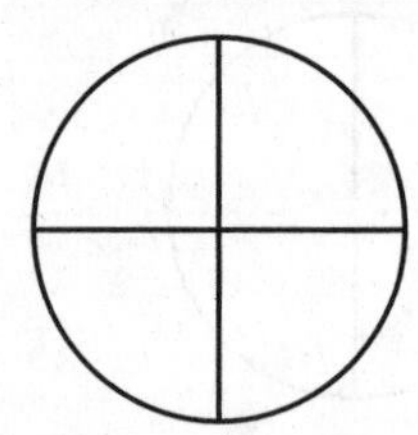

____

Shade three eighths. Write the fraction.

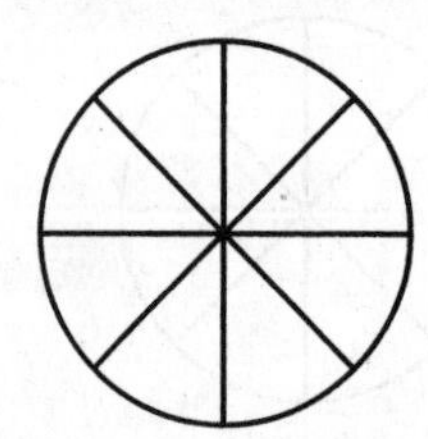

____

Shade one half. Write the fraction.

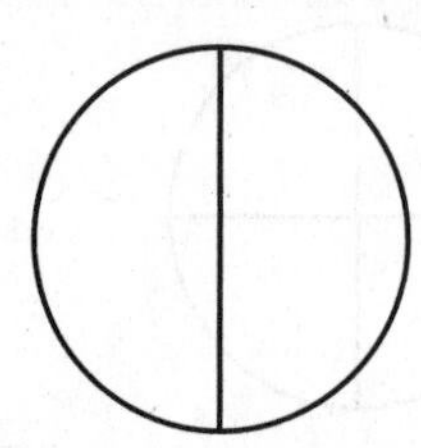

____

4. Make 27¢ using the fewest dimes and pennies.

_______ dimes _______ pennies

Make 45 using the fewest tens and ones.

_______ tens _______ ones

5. Count by 10's. 10, 20, 30, _______, _______, _______, _______, _______

6. Find the sums.

70 + 10 = _____ 20 + 10 = _____ 10 + 90 = _____

Name ____________________

Class Fact Practice 42A

***Saxon** Math 2 (for use with **Lesson 42**)*

Set 8: Sums of 11; Review Facts

$$\begin{array}{r} 2 \\ +\ 9 \\ \hline \end{array} \quad \begin{array}{r} 3 \\ +\ 7 \\ \hline \end{array} \quad \begin{array}{r} 4 \\ +\ 3 \\ \hline \end{array} \quad \begin{array}{r} 9 \\ +\ 8 \\ \hline \end{array} \quad \begin{array}{r} 6 \\ +\ 5 \\ \hline \end{array}$$

$$\begin{array}{r} 4 \\ +\ 6 \\ \hline \end{array} \quad \begin{array}{r} 5 \\ +\ 9 \\ \hline \end{array} \quad \begin{array}{r} 7 \\ +\ 4 \\ \hline \end{array} \quad \begin{array}{r} 3 \\ +\ 5 \\ \hline \end{array} \quad \begin{array}{r} 9 \\ +\ 2 \\ \hline \end{array}$$

$$\begin{array}{r} 2 \\ +\ 5 \\ \hline \end{array} \quad \begin{array}{r} 7 \\ +\ 9 \\ \hline \end{array} \quad \begin{array}{r} 6 \\ +\ 4 \\ \hline \end{array} \quad \begin{array}{r} 8 \\ +\ 3 \\ \hline \end{array} \quad \begin{array}{r} 9 \\ +\ 6 \\ \hline \end{array}$$

$$\begin{array}{r} 4 \\ +\ 7 \\ \hline \end{array} \quad \begin{array}{r} 3 \\ +\ 4 \\ \hline \end{array} \quad \begin{array}{r} 6 \\ +\ 3 \\ \hline \end{array} \quad \begin{array}{r} 9 \\ +\ 4 \\ \hline \end{array} \quad \begin{array}{r} 5 \\ +\ 6 \\ \hline \end{array}$$

$$\begin{array}{r} 7 \\ +\ 3 \\ \hline \end{array} \quad \begin{array}{r} 4 \\ +\ 5 \\ \hline \end{array} \quad \begin{array}{r} 3 \\ +\ 8 \\ \hline \end{array} \quad \begin{array}{r} 6 \\ +\ 2 \\ \hline \end{array} \quad \begin{array}{r} 3 \\ +\ 9 \\ \hline \end{array}$$

Name ____________________

Saxon Math 2 (for use with Lesson 42)

Set 8: Sums of 11; Review Facts

1. Read the answers to someone.
2. Write the answers.
3. Ask someone to correct your paper. Corrected by ____________________

5 + 6	9 + 4	6 + 3	3 + 4	4 + 7
9 + 6	8 + 3	6 + 4	7 + 9	2 + 5
9 + 2	3 + 5	7 + 4	5 + 9	4 + 6
3 + 9	6 + 2	3 + 8	4 + 5	7 + 3
6 + 5	9 + 8	4 + 3	3 + 7	2 + 9

Name ______________________

Date ______________________

1. Marcus had 4 dimes. His mother gave him 1 dime. How many dimes does Marcus have now? Draw a picture and write a number sentence for this story. Write the answer with a label.

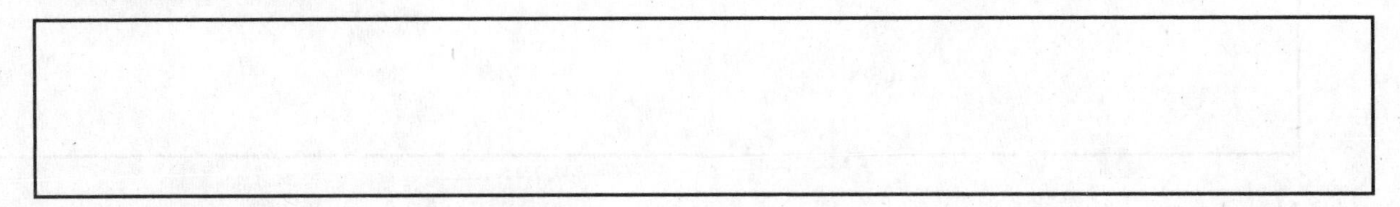

Number sentence ______________________

Answer ______________________

How much money is this? __________

2. Shade one fourth. Write the fraction.

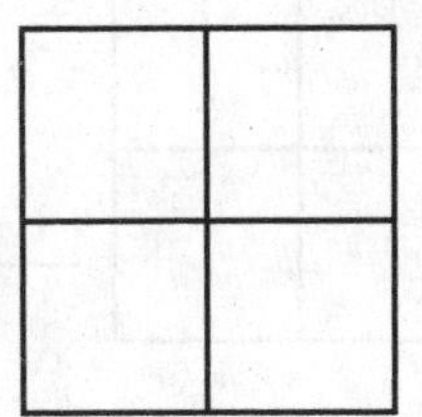

Shade three sixths. Write the fraction.

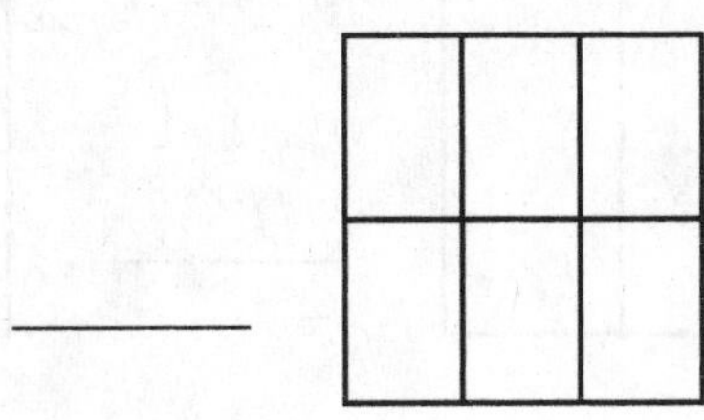

Shade two eighths. Write the fraction.

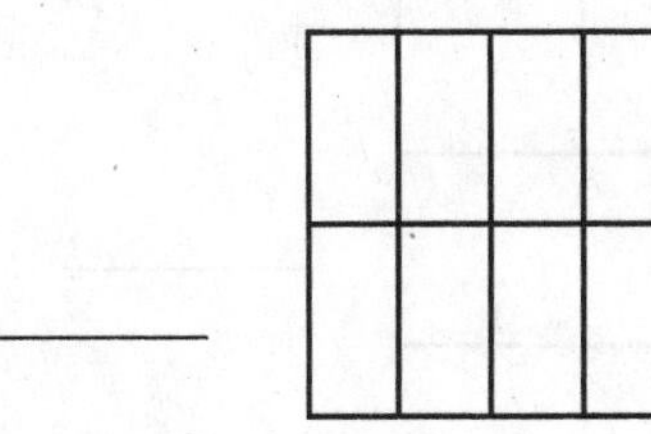

3. Use your classroom's favorite apples graph to answer these questions.

How many children voted? ________

What was the favorite apple? ______________

How many children chose that apple? ________

4. I have 3 dimes and 4 pennies. How much money is that? __________

I have 5 tens and 6 ones. How much is that? ________

Make 23¢ using the fewest dimes and pennies.

________ dimes ________ pennies

5. Fill in the missing addends.

$4 + \square = 10$ $\square + 2 = 10$ $\square + 7 = 10$

Name ______________________________

Homework 42B

***Saxon** Math 2 (for use with **Lesson 42**)*

1. Althea had 6 dimes. She spent 1 dime. How many dimes does she have now? Draw a picture and write a number sentence for this story. Write the answer with a label.

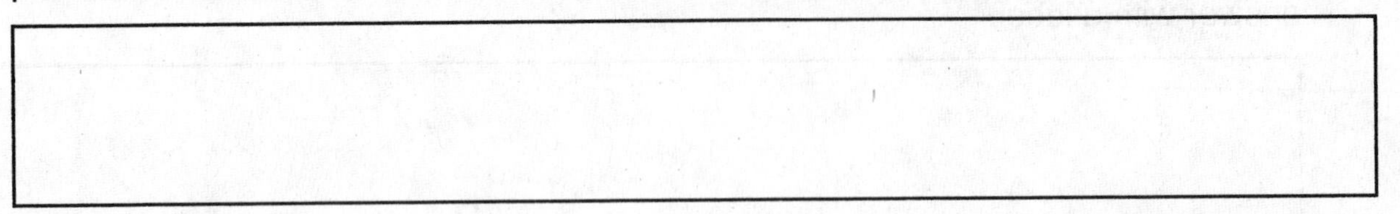

Number sentence ______________________________

Answer ______________________

How much money is this? __________

2. Shade three fourths. Write the fraction.

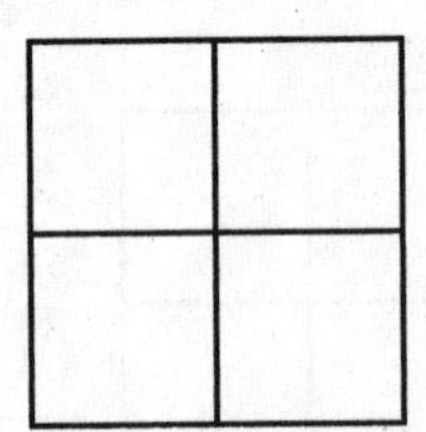

______

Shade one half. Write the fraction.

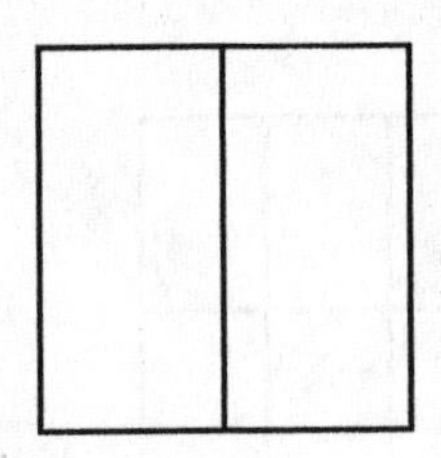

______

Shade six eighths. Write the fraction.

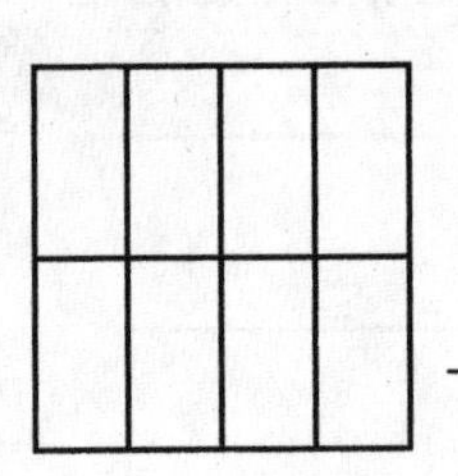

______

3. Fill in the missing numbers in these number patterns.

5, 10, 15, ______, ______, ______, ______, ______, ______, ______

10, 20, 30, ______, ______, ______, ______, ______, ______, ______

4. I have 2 dimes and 5 pennies. How much money is that? ______

I have 9 tens and 7 ones. How much is that? ______

Make 74¢ using the fewest dimes and pennies.

______ dimes ______ pennies

5. Fill in the missing addends.

☐ + 3 = 10 6 + ☐ = 10 ☐ + 8 = 10

Name ______________________________

Class Fact Practice 43A

***Saxon** Math 2 (for use with **Lesson 43**)*

Set 8: Sums of 11; Review Facts

A.

7 + 3 = ____	9 + 2 = ____	5 + 9 = ____
4 + 5 = ____	6 + 4 = ____	3 + 8 = ____
4 + 3 = ____	4 + 7 = ____	9 + 8 = ____
2 + 9 = ____	3 + 5 = ____	5 + 6 = ____
9 + 4 = ____	8 + 3 = ____	7 + 9 = ____
3 + 7 = ____	9 + 6 = ____	7 + 4 = ____
3 + 9 = ____	6 + 5 = ____	4 + 6 = ____

B.

2 + ☐ = 11	8 + ☐ = 11	4 + ☐ = 11
☐ + 7 = 11	☐ + 9 = 11	☐ + 3 = 11

Name ______________________

Fact Homework 43B

***Saxon** Math 2 (for use with **Lesson 43**)*

Set 8: Sums of 11; Review Facts

A. 1. Read the answers to someone.
2. Write the answers.
3. Ask someone to correct your paper. Corrected by ______________

$4 + 6 =$ ____	$6 + 5 =$ ____	$3 + 9 =$ ____
$7 + 4 =$ ____	$9 + 6 =$ ____	$3 + 7 =$ ____
$7 + 9 =$ ____	$8 + 3 =$ ____	$9 + 4 =$ ____
$5 + 6 =$ ____	$3 + 5 =$ ____	$2 + 9 =$ ____
$9 + 8 =$ ____	$4 + 7 =$ ____	$4 + 3 =$ ____
$3 + 8 =$ ____	$6 + 4 =$ ____	$4 + 5 =$ ____
$5 + 9 =$ ____	$9 + 2 =$ ____	$7 + 3 =$ ____

B. Fill in the missing numbers.

$7 + \square = 11$	$9 + \square = 11$	$3 + \square = 11$
$\square + 2 = 11$	$\square + 8 = 11$	$\square + 4 = 11$

Name ________________________________

Measure the line segments to the nearest inch.

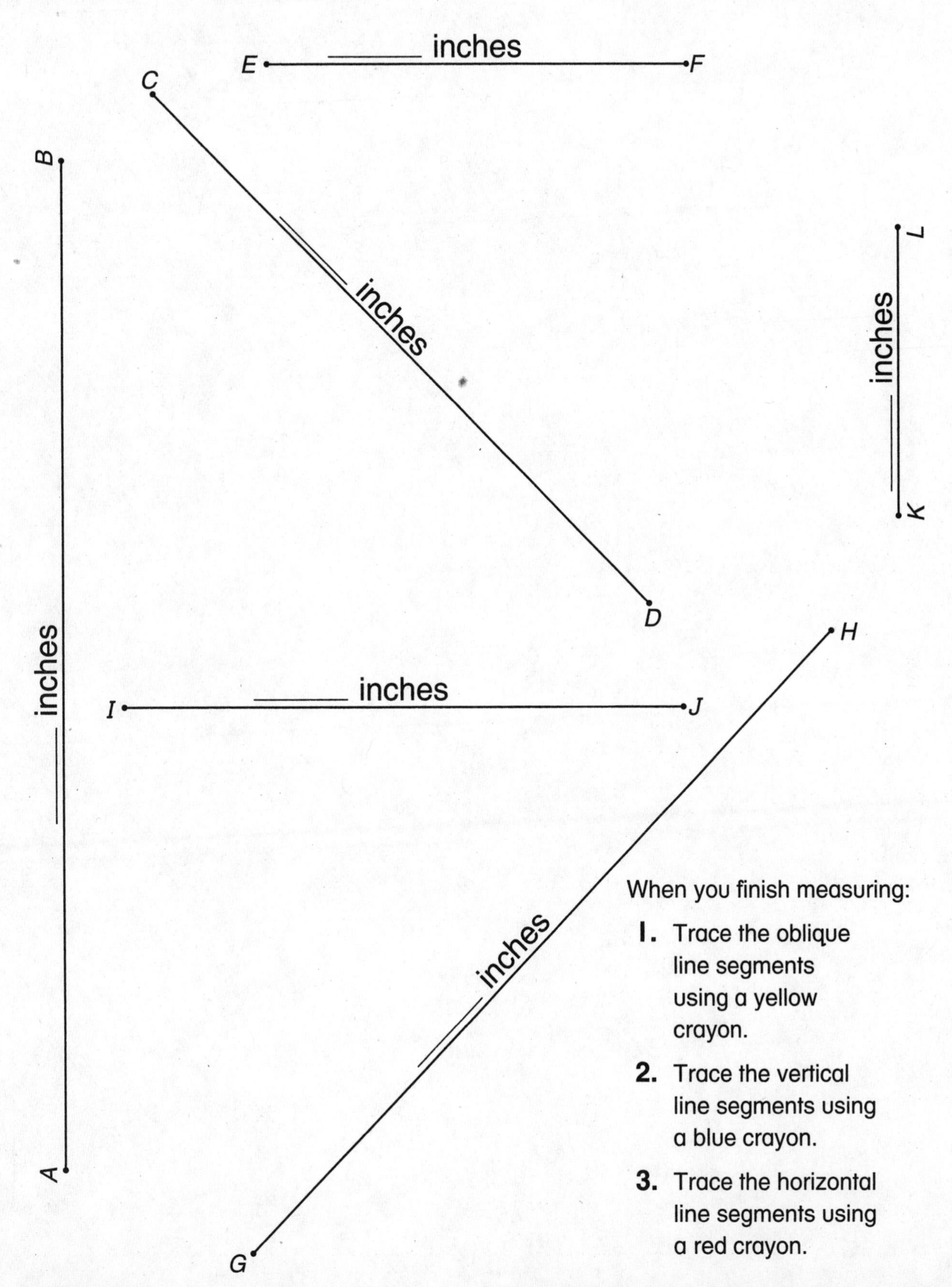

When you finish measuring:

1. Trace the oblique line segments using a yellow crayon.
2. Trace the vertical line segments using a blue crayon.
3. Trace the horizontal line segments using a red crayon.

Name ______________________

Date ______________________

Guided Class Practice 43A

Saxon Math 2 (for use with Lesson 43)

1. On Fridays the children can buy pizza or hot dogs for lunch. Last Friday eighty children bought pizza and ten children bought hot dogs. How many children bought lunch last Friday?

Number sentence ______________________

Answer ______________________

2. Divide the square in half using a vertical line segment.

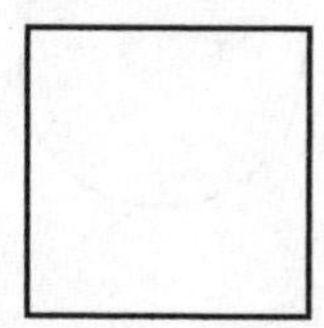

Divide the square in half using a horizontal line segment.

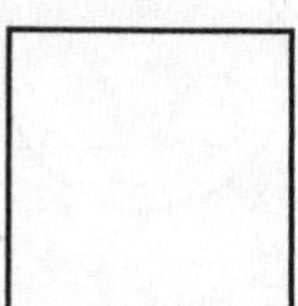

Divide the square in half using an oblique line segment.

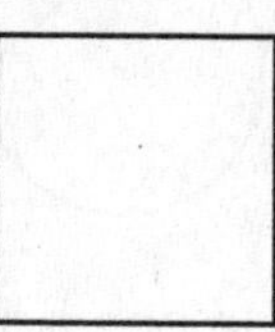

Shade one half of each square.

3. How much money is 2 dimes and 6 pennies? ________

How much money is 2 pennies and 6 dimes? ________

Circle the one that is more.

4. Measure this line segment using inches.

•——————————————————————• ________"

5. Fill in the missing numbers in this fact family.

3 + ☐ = 11   8 + ☐ = 11

11 − ☐ = 3   ☐ − 3 = 8

6. Count by 5's.

5, ____, ____, ____, ____, ____, ____, ____, ____, ____

Name ______________________________

Date ______________________________

Homework 43B

***Saxon** Math 2 (for use with **Lesson 43**)*

1. Crystal read the first 40 pages of the book on Saturday. On Sunday she read 10 more pages. How many pages did she read altogether?

Number sentence ______________________________

Answer ______________________________

2. Divide the circle in half using a vertical line segment.

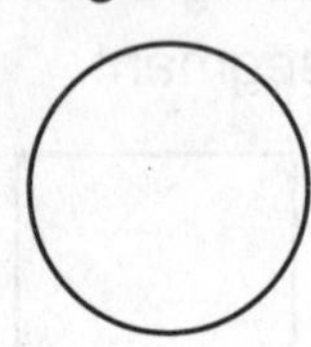

Divide the circle in half using a horizontal line segment.

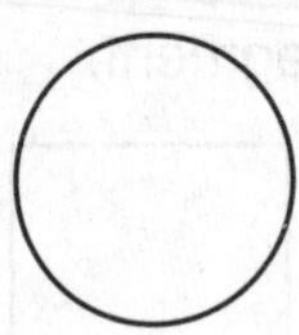

Divide the circle in half using an oblique line segment.

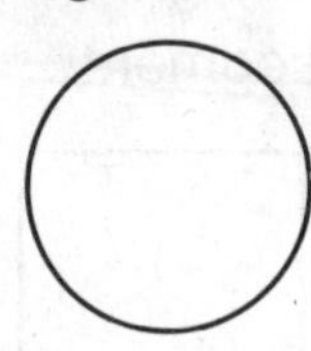

Shade one half of each circle.

3. How much money is 3 dimes and 5 pennies? ____________

How much money is 3 pennies and 5 dimes? ____________

Circle the one that is more.

4. Find the sums.

80 + 10 = ______ 10 + 40 = ______ 70 + 10 = ______

5. Fill in the missing numbers in this fact family.

4 + ☐ = 11 7 + ☐ = 11

11 − ☐ = 7 ☐ − 7 = 4

6. Count by 10's.

10, ______, ______, ______, ______, ______, ______, ______, ______, ______

Name ______________________________

Class Fact Practice 44A

***Saxon** Math 2 (for use with **Lesson 44**)*

Set 8: Sums of 11; Review Facts

3 + 4	6 + 5	7 + 9	4 + 6	8 + 3
9 + 6	2 + 9	3 + 5	4 + 7	9 + 8
3 + 9	6 + 2	5 + 6	7 + 3	5 + 9
7 + 4	6 + 3	9 + 4	3 + 8	4 + 3
6 + 4	2 + 5	9 + 2	4 + 5	3 + 7

Name ____________________

Fact Homework 44B

***Saxon** Math 2 (for use with **Lesson 44**)*

Set 8: Sums of 11; Review Facts

1. Read the answers to someone.
2. Write the answers.
3. Ask someone to correct your paper. Corrected by ____________________

$4 + 3$	$3 + 8$	$9 + 4$	$6 + 3$	$7 + 4$
$3 + 7$	$4 + 5$	$9 + 2$	$2 + 5$	$6 + 4$
$8 + 3$	$4 + 6$	$7 + 9$	$6 + 5$	$3 + 4$
$9 + 8$	$4 + 7$	$3 + 5$	$2 + 9$	$9 + 6$
$5 + 9$	$7 + 3$	$5 + 6$	$6 + 2$	$3 + 9$

Name ______________________________

Date ______________________________

Guided Class Practice 44A

***Saxon** Math 2 (for use with **Lesson 44**)*

1. Stuart has 10 large marbles and 24 small marbles. How many marbles does he have?

Number sentence ______________________________

Answer ______________________

2. Measure this line segment using inches.

•————————————————————————• ______ "

3. Show 57¢ using the fewest dimes and pennies.

______ dimes ______ pennies

4. 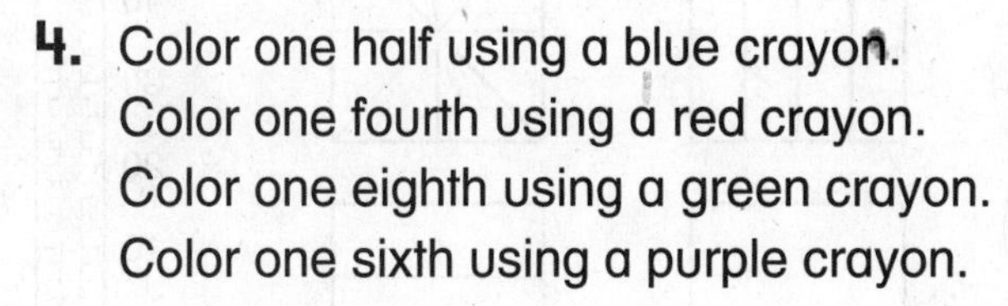

Color one half using a blue crayon.
Color one fourth using a red crayon.
Color one eighth using a green crayon.
Color one sixth using a purple crayon.

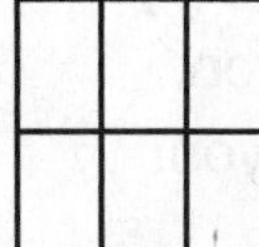 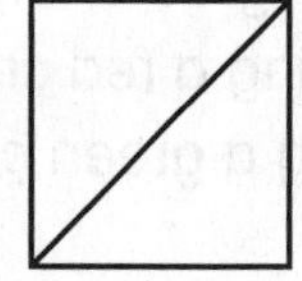 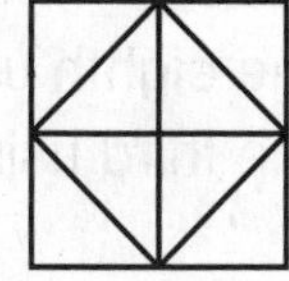 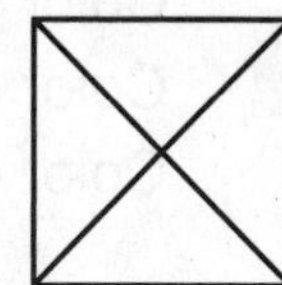

5. Find the sums.

36 + 10 = ______ 28 + 10 = ______

10 + 42 = ______ 73 + 10 = ______

6. Fill in the missing numbers on this piece of a hundred number chart.

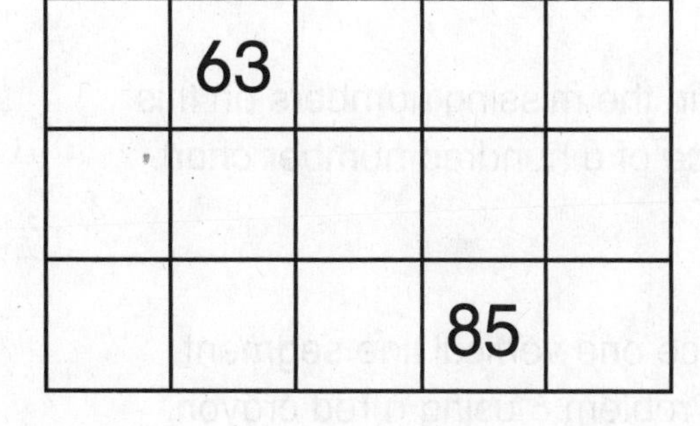

7. Trace one horizontal line segment in Problem 6 using a red crayon.

Name ______________________________

Date ______________________________

Homework 44B

***Saxon** Math 2 (for use with **Lesson 44**)*

1. Clara had 35 pennies. Her brother gave her 10 more pennies. How many pennies does she have now?

Number sentence ______________________________

Answer ______________________

2. Which number on the thermometer is the temperature closest to? ________°F

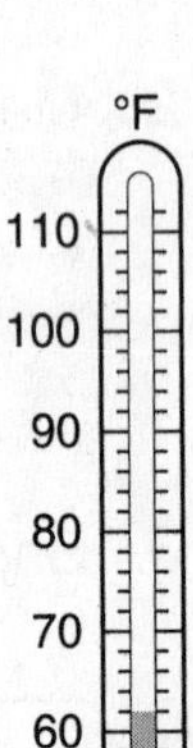

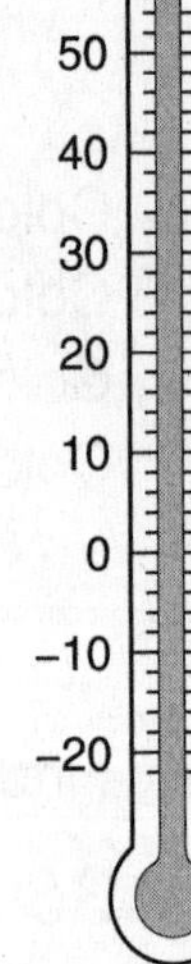

3. Show 87¢ using the fewest dimes and pennies.

________ dimes ________ pennies

4. Color one half using a blue crayon.
Color one fourth using an orange crayon.
Color one eighth using a red crayon.
Color one third using a green crayon.

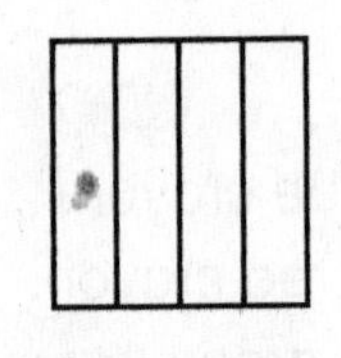
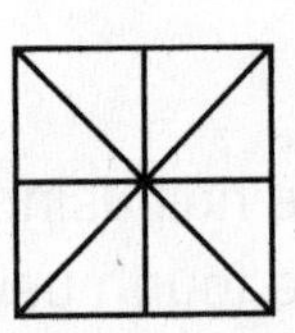
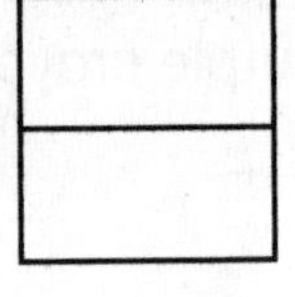
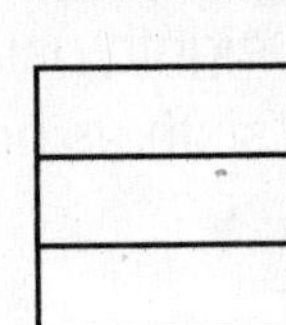

5. Find the sums.

$85 + 10 =$ ______

$29 + 10 =$ ______

$10 + 37 =$ ______

$10 + 21 =$ ______

6. Fill in the missing numbers on this piece of a hundred number chart.

7. Trace one vertical line segment in Problem 6 using a red crayon.

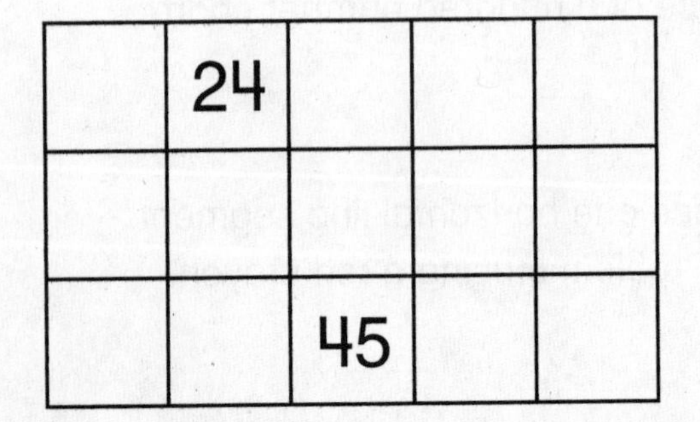

Name ____________________ Score ________

Fact Assessment 8

***Saxon** Math 2 (for use with **Lesson 45-1**)*

Set 8: Sums of 11; Review Facts

$6 + 5$	$9 + 8$	$2 + 9$	$6 + 4$	$4 + 5$
$8 + 3$	$5 + 9$	$3 + 7$	$9 + 2$	$4 + 3$
$6 + 2$	$4 + 7$	$3 + 5$	$9 + 4$	$3 + 8$
$7 + 9$	$4 + 6$	$3 + 9$	$6 + 3$	$7 + 4$
$9 + 6$	$3 + 4$	$5 + 6$	$7 + 3$	$2 + 5$

Name ______________________

Lesson Worksheet 45-1A

Saxon Math 2 (for use with Lesson 45-1)

A.

$3 + \square = 12$ $4 + \square = 12$ $5 + \square = 12$ $6 + \square = 12$ $7 + \square = 12$

$9 + \square = 12$ $8 + \square = 12$ $7 + \square = 12$ $6 + \square = 12$ $5 + \square = 12$

B.

$5 + \square = 12$ $9 + \square = 12$ $7 + \square = 12$ $3 + \square = 12$ $8 + \square = 12$

$6 + \square = 12$ $4 + \square = 12$ $\square + 3 = 12$ $\square + 5 = 12$ $\square + 8 = 12$

$\square + 9 = 12$ $\square + 4 = 12$ $\square + 7 = 12$ $\square + 6 = 12$ $\square + 3 = 12$

Dear Parent,

Today we practiced the sum of 12 facts. Your child will be tested next week on the new addition facts listed below, along with previously introduced facts.

$$\begin{array}{r} 4 \\ +\ 8 \\ \hline \end{array} \qquad \begin{array}{r} 8 \\ +\ 4 \\ \hline \end{array} \qquad \begin{array}{r} 5 \\ +\ 7 \\ \hline \end{array} \qquad \begin{array}{r} 7 \\ +\ 5 \\ \hline \end{array}$$

During the next four days, your child will practice these facts in class both orally and in writing. The back of each day's fact sheet will contain the fact practice homework.

Keep this paper at home to use for practice.

Name ______________________

Date ______________________

1. Jay counted nine sharpened pencils and six unsharpened pencils in the pencil can. How many pencils are in the can?

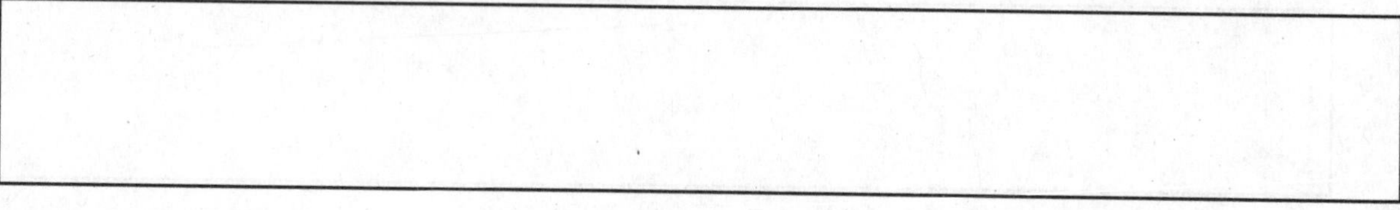

Number sentence ______________________

Answer ______________________

2. Which number on the thermometer is the temperature closest to? ________ °F

3. How many mittens are in the box? ________ mittens

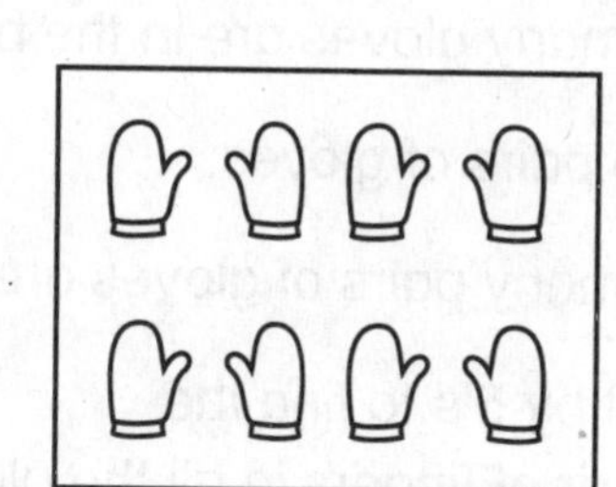

Circle pairs of mittens.

How many pairs of mittens are there? ________ pairs

Count by 5's to find the number of fingers in all the mittens. ________ fingers

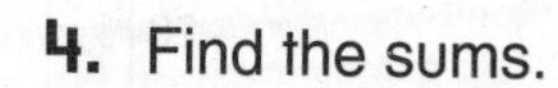

4. Find the sums.

56 + 10 = ________ 10 + 35 = ________

5. How many dimes are there? ________

How many pennies are there? ________

How much money is this? ________

Name ______________________

Date ______________________

Homework 45B

***Saxon** Math 2 (for use with **Lesson 45-1**)*

1. Ferma has a set of 10 markers. She threw away 3 markers because they were dry. How many markers does she have now?

Number sentence ______________________

Answer ______________________

2. Which number on the thermometer is the temperature closest to? ________ °F

°F
110
100
90
80
70
60
50
40
30
20
10
0
–10
–20

3. How many gloves are in the box? ________ gloves

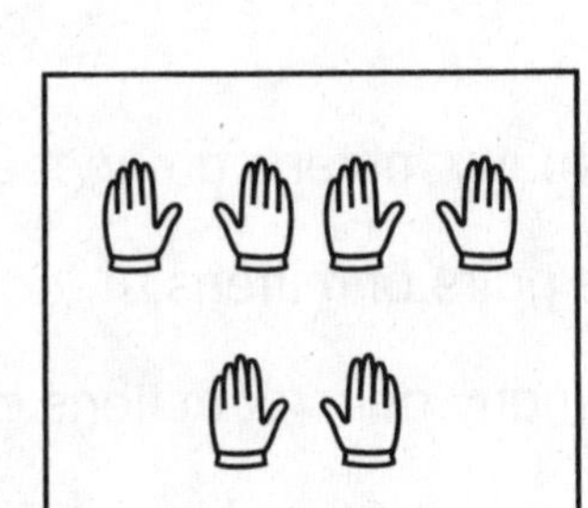

Circle pairs of gloves.

How many pairs of gloves are there? ________ pairs

Count by 5's to find the number of fingers in all the gloves. ________ fingers

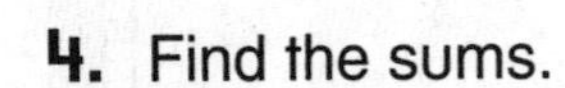

4. Find the sums.

$39 + 10 =$ ________ $10 + 62 =$ ________

5. How many dimes are there? ________

How many pennies are there? ________

How much money is this? ________

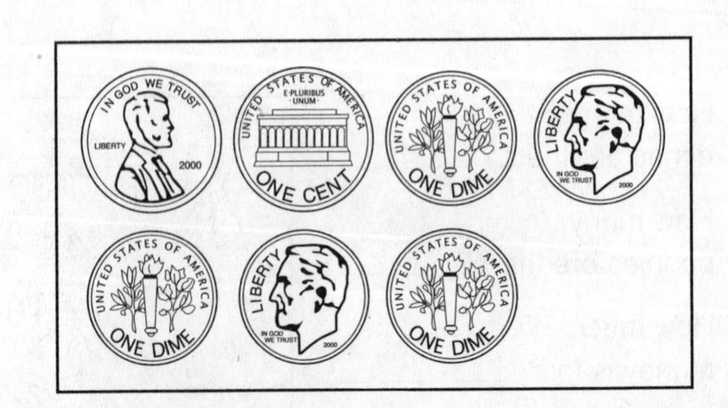

Name ______________________

Date ______________________

Written Assessment 8

***Saxon** Math 2 (for use with **Lesson 45-2**)*

1. Steven has four dimes. Mark has two dimes. How many dimes do they have altogether? Draw a picture and write a number sentence for this story. Write the answer with a label.

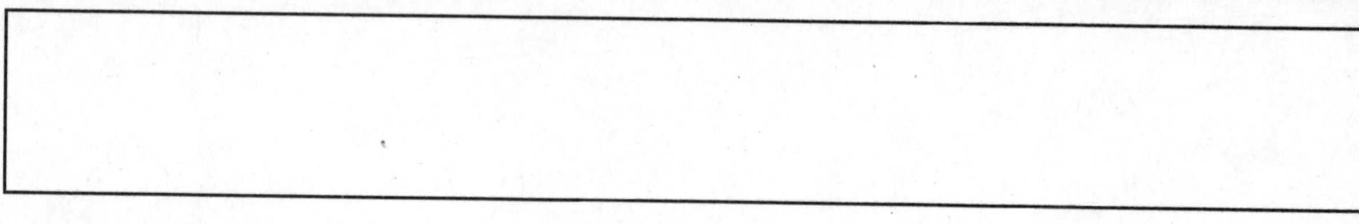

Number sentence ______________________

Answer ______________________

2. I have 7 dimes and 2 pennies. How much money is that? ________

I have 3 tens and 8 ones. How much is that? ______

3. Count by 10's.

10, ____, ____, ____, ____, ____, ____, ____, ____, ____

Count by 5's.

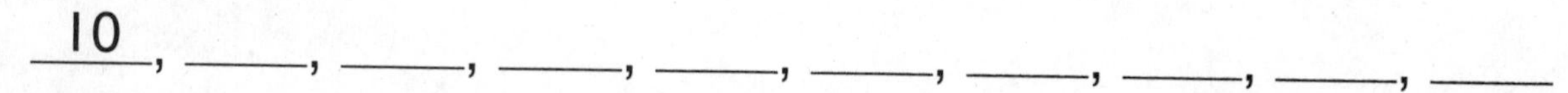

5, ____, ____, ____, ____, ____, ____, ____, ____, ____

4. Color the circle divided into fourths red.
Color the circle divided into halves blue.
Color the circle divided into eighths green.

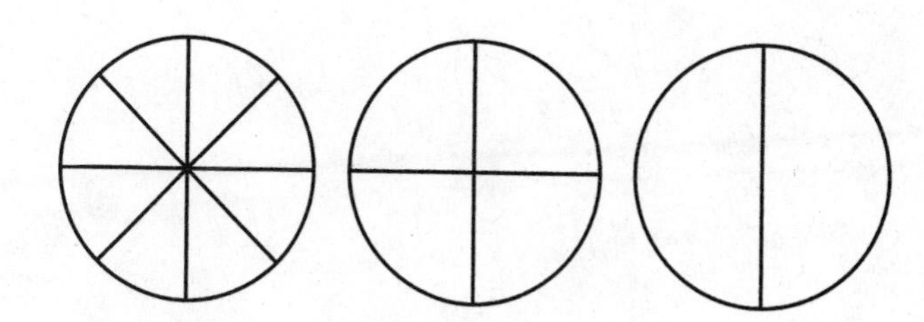

5. Find the sums.

40 + 10 = ____ 10 + 80 = ____ 10 + 70 = ____

6. Fill in the missing addends.

☐ + 7 = 10 4 + ☐ = 10 ☐ + 2 = 10

Name ______________________________

Class Fact Practice 46A

***Saxon** Math 2 (for use with **Lesson 46**)*

Set 9: Sums of 12; Review Facts

$3 + 9$	$5 + 4$	$6 + 7$	$8 + 4$	$6 + 1$
$4 + 7$	$1 + 8$	$7 + 5$	$9 + 2$	$6 + 4$
$3 + 8$	$2 + 4$	$6 + 6$	$3 + 7$	$9 + 5$
$4 + 8$	$6 + 9$	$2 + 8$	$9 + 3$	$5 + 2$
$8 + 3$	$5 + 7$	$8 + 7$	$7 + 4$	$3 + 6$

Name ________________________________

Fact Homework 46B

***Saxon** Math 2 (for use with **Lesson 46**)*

Set 9: Sums of 12; Review Facts

1. Read the answers to someone.
2. Write the answers.
3. Ask someone to correct your paper. Corrected by ____________________

3 + 6	7 + 4	8 + 7	5 + 7	8 + 3
5 + 2	9 + 3	2 + 8	6 + 9	4 + 8
9 + 5	3 + 7	6 + 6	2 + 4	3 + 8
6 + 4	9 + 2	7 + 5	1 + 8	4 + 7
6 + 1	8 + 4	6 + 7	5 + 4	3 + 9

Name ______________________

Date ______________________

Guided Class Practice 46A

***Saxon** Math 2 (for use with **Lesson 46**)*

1. The children in Room 12 collected 83 cans of food for the food drive. Ms. Roman brought ten more cans of food. How many cans of food did they collect altogether?

Number sentence ______________________

Answer ______________________

2. Find each sum. $10 + 36 =$ ______ $13 + 10 =$ ______

3. Measure each line segment using inches.

How long is the horizontal line segment? ______"

How long is the vertical line segment? ______"

4. Chris has nine white socks.
Draw the socks and circle the pairs.

How many pairs are there? ______ pairs

How many extras are there? ______ extra

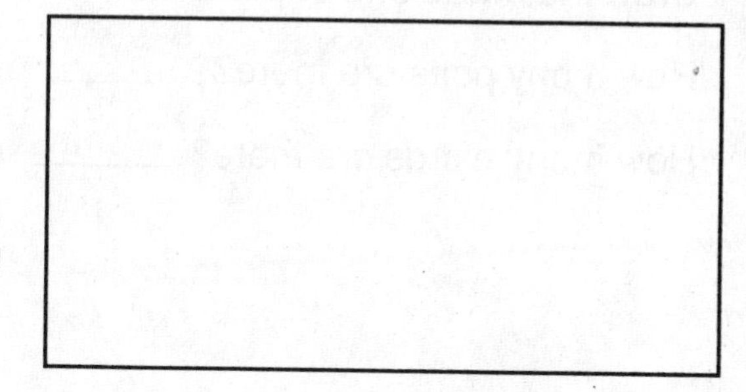

5. Write the even and odd numbers.

0, 2, 4, ____, ____, ____, ____, ____, ____, ____, ____

1, 3, 5, ____, ____, ____, ____, ____, ____, ____, ____

6. How much money is this? ______________

Name ______________________

Date ______________________

Homework 46B

***Saxon** Math 2 (for use with **Lesson 46**)*

1. Ten children in Room 17 voted yes and 31 children voted no. How many children voted altogether?

Number sentence ______________________

Answer ______________________

2. Find each sum. 10 + 43 = ______ 16 + 10 = ______

3. Color the fifth triangle blue.
Color the sixth triangle red.
Color the middle triangle yellow.

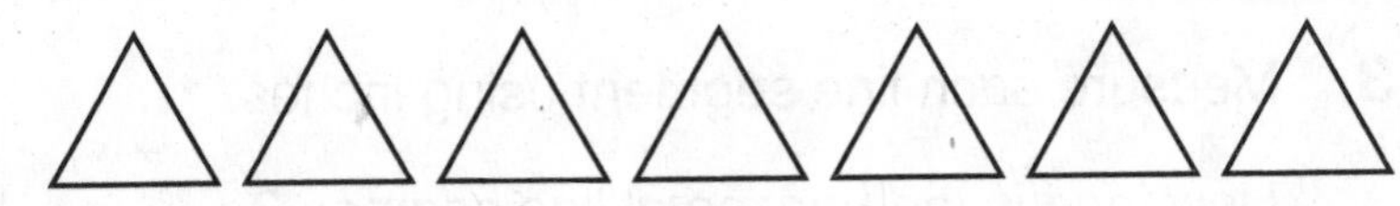

4. Missy has seven white socks.
Draw the socks and circle the pairs.

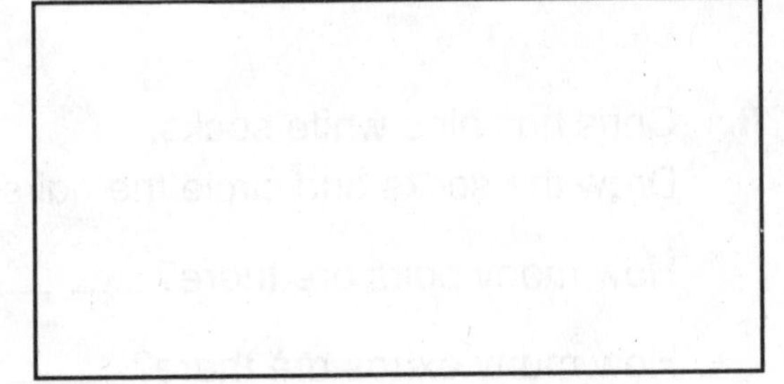

How many pairs are there? ______ pairs

How many extras are there? ______ extra

5. Color the even numbers red. Color the odd numbers yellow.

1	2	3	4	5	6	7	8	9	10
11	12	13	14	15	16	17	18	19	20

6. How much money is this? ______

Name ______________________

Class Fact Practice 47A

***Saxon** Math 2 (for use with **Lesson 47**)*

Set 9: Sums of 12; Review Facts

$6 + 4$	$9 + 2$	$7 + 5$	$2 + 8$	$9 + 5$
$3 + 8$	$6 + 6$	$2 + 4$	$3 + 9$	$7 + 4$
$8 + 7$	$5 + 2$	$4 + 7$	$1 + 8$	$8 + 4$
$6 + 7$	$9 + 3$	$3 + 6$	$5 + 7$	$8 + 3$
$5 + 4$	$3 + 7$	$4 + 8$	$6 + 1$	$6 + 9$

Name ______________________________

Set 9: Sums of 12; Review Facts

1. Read the answers to someone.
2. Write the answers.
3. Ask someone to correct your paper. Corrected by ______________________

$9 + 3$	$2 + 4$	$5 + 7$	$1 + 8$	$8 + 7$
$9 + 5$	$7 + 4$	$5 + 2$	$4 + 8$	$3 + 7$
$5 + 4$	$3 + 9$	$6 + 4$	$6 + 1$	$3 + 8$
$8 + 4$	$3 + 6$	$2 + 8$	$7 + 5$	$9 + 2$
$6 + 6$	$4 + 7$	$6 + 9$	$8 + 3$	$6 + 7$

Name ______________________________

Date ______________________________

Guided Class Practice 47A

***Saxon** Math 2 (for use with **Lesson 47**)*

1. Corrine has 5 dogs, 2 cats, 3 birds, and a turtle as pets. How many of her pets have fur?

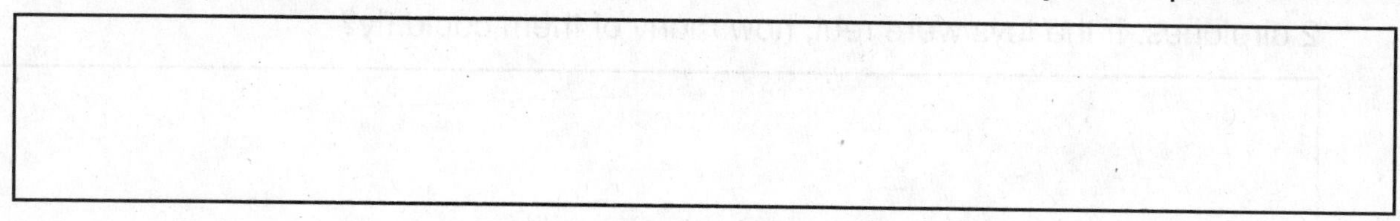

Number sentence ______________________________

Answer ______________________

2. Finish the patterns.

2, 4, 6, ____, ____, ____, ____, ____, ____, ____

5, 10, 15, ____, ____, ____, ____, ____, ____, ____

3. Write these dates using digits. June 20, 2006 __________

November 4, 2011 __________

4. How much money is this? __________

5. Make 82¢ using the fewest dimes and pennies.

______ dimes ______ pennies

Make 37 using the fewest tens and ones.

______ tens ______ ones

6. Fill in the missing addends.

$6 + \square = 10$ 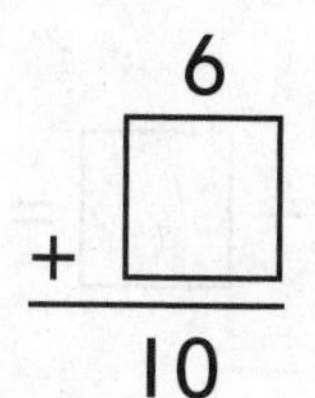

$\square + 2 = 7$

$1 + \square = 6$

$\square + 3 = 7$

$7 + \square = 10$

$\square + 2 = 8$

Name ____________________

Date ____________________

Homework 47B

Saxon Math 2 (for use with **Lesson 47**)

1. Chris has a collection of 3 helicopters, 1 fire engine, 2 police cars, 1 motorcycle, and 2 airplanes. If the toys were real, how many of them could fly?

Number sentence ____________________

Answer ____________________

2. Finish the patterns.

1, 3, 5, ____, ____, ____, ____, ____, ____, ____

10, 20, 30, ____, ____, ____, ____, ____, ____, ____

3. Write these dates using digits. January 15, 2007 ____________

July 4, 2024 ____________

4. How much money is this? ____________

5. Make 57¢ using the fewest dimes and pennies.

______ dimes ______ pennies

Make 29 using the fewest tens and ones.

______ tens ______ ones

6. Fill in the missing addends.

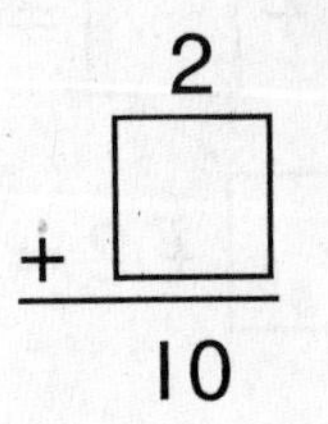

$2 + \square = 10$

$\square + 1 = 9$

$4 + \square = 8$

$\square + 7 = 13$

$3 + \square = 10$

$\square + 2 = 6$

Name ______________________________

Set 9: Sums of 12; Review Facts

A.

6 + 7 = ____	3 + 9 = ____	6 + 4 = ____
5 + 4 = ____	8 + 3 = ____	5 + 7 = ____
9 + 2 = ____	9 + 5 = ____	8 + 4 = ____
5 + 2 = ____	3 + 8 = ____	8 + 7 = ____
4 + 7 = ____	2 + 8 = ____	9 + 3 = ____
3 + 7 = ____	6 + 9 = ____	7 + 4 = ____
4 + 8 = ____	3 + 6 = ____	7 + 5 = ____

B.

5 + ☐ = 12	9 + ☐ = 12	4 + ☐ = 12
☐ + 8 = 12	☐ + 7 = 12	☐ + 3 = 12

Name ______________________

Set 9: Sums of 12; Review Facts

A. 1. Read the answers to someone.
2. Write the answers.
3. Ask someone to correct your paper. Corrected by ______________________

9 + 3 = ____ 2 + 8 = ____ 4 + 7 = ____

8 + 7 = ____ 3 + 8 = ____ 5 + 2 = ____

8 + 4 = ____ 9 + 5 = ____ 9 + 2 = ____

5 + 7 = ____ 8 + 3 = ____ 5 + 4 = ____

6 + 4 = ____ 3 + 9 = ____ 6 + 7 = ____

7 + 5 = ____ 3 + 6 = ____ 4 + 8 = ____

7 + 4 = ____ 6 + 9 = ____ 3 + 7 = ____

B. Fill in the missing numbers.

8 + ☐ = 12 3 + ☐ = 12 5 + ☐ = 12

☐ + 4 = 12 ☐ + 7 = 12 ☐ + 9 = 12

Name ______________________________

Date ______________________________

1. Susan had 12 pencils. Marsha gave Susan two pencils. How many pencils does Susan have now?

Number sentence ______________________________

Answer ______________________________

2. Underline all the circles that are divided correctly into fourths (four equal parts).

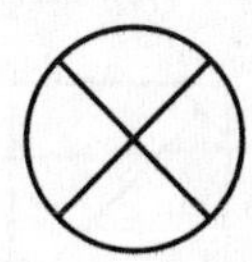 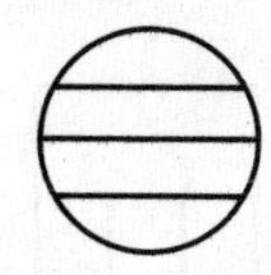 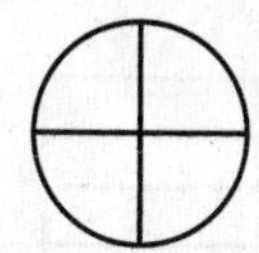 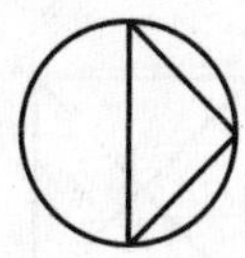

3. Show **18** using tally marks.

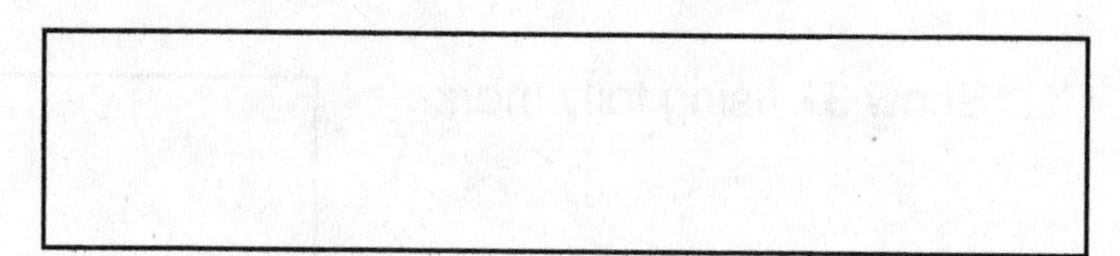

4. One of these socks is my favorite sock.
Use the clues to find my favorite sock.
Cross out the socks that cannot be my favorite sock.

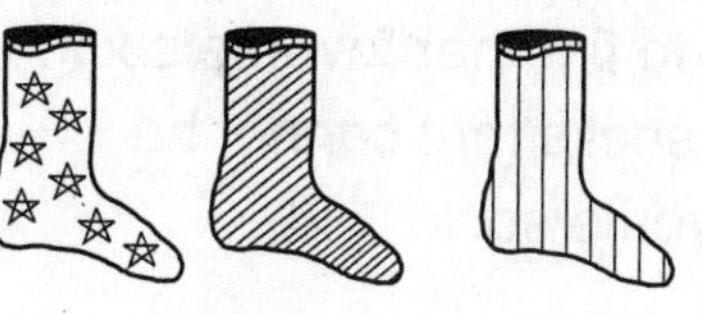
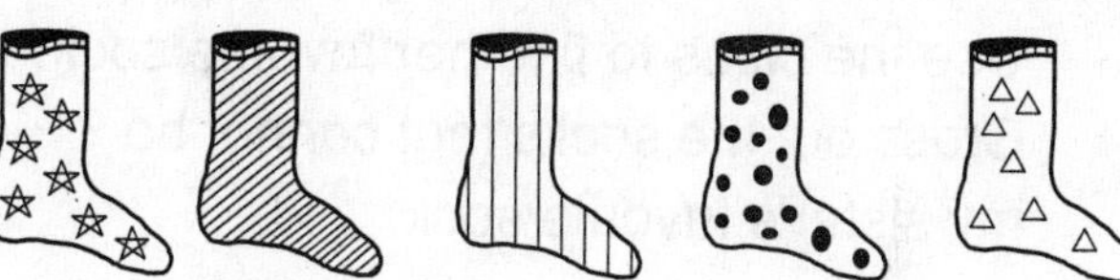

It is not the sock with the triangles.
It is not the sock with the vertical lines.
It is not first.
It is not fourth.
Circle my favorite sock.

5. Six children in Miss Wood's class graphed their tags.

How many families have boys? ________

How many families have only girls? ________

How many families have both boys and girls? ________

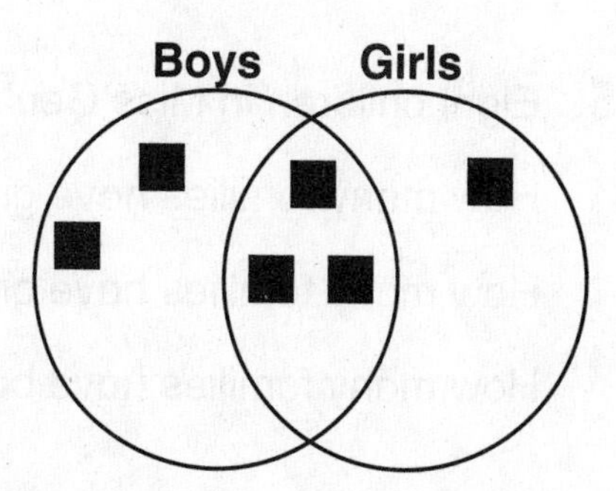

6. Find the sums.

$18 + 10 =$ ________ $77 + 10 =$ ________

$10 + 52 =$ ________ $10 + 37 =$ ________

Name ______________________

Date ______________________

Homework 48B

***Saxon** Math 2 (for use with **Lesson 48**)*

1. Mrs. Murray had 6 games. Mrs. Murray gave Mrs. Paolino's class 3 games to use. How many games does Mrs. Murray have now?

Number sentence ______________________

Answer ______________________

2. Circle all the squares that are divided correctly into fourths (four equal parts).

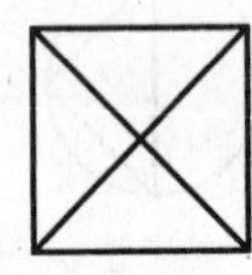 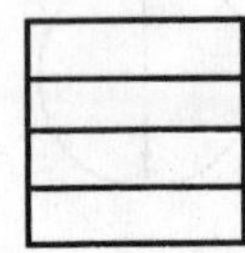 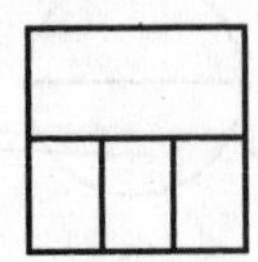 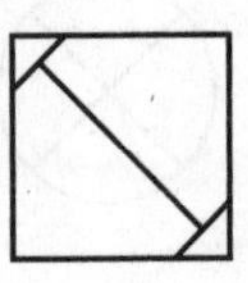

3. Show **31** using tally marks.

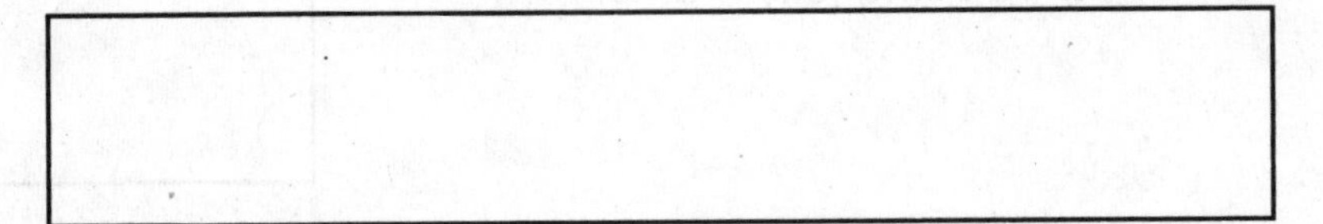

4. One of these socks is my sister's favorite sock. Use the clues to find her favorite sock. Cross out the socks that cannot be my sister's favorite sock.

It is not the sock with the triangles.
It is not the sock with the vertical lines.
It is not first.
It is not second.
Circle my sister's favorite sock.

5. Eight children in Miss Gen's class graphed their tags.

How many families have girls? ________

How many families have only boys? ________

How many families have both boys and girls? ________

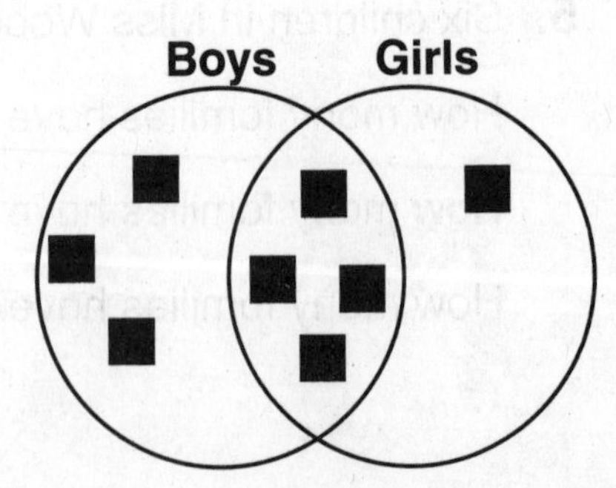

6. Find the sums.

26 + 10 = ________ 86 + 10 = ________

10 + 32 = ________ 10 + 68 = ________

Name ______________________

Saxon Math 2 (for use with Lesson 49)

Set 9: Sums of 12; Review Facts

9 + 2	7 + 5	2 + 8	3 + 6	8 + 4
6 + 7	8 + 3	6 + 9	4 + 7	6 + 6
8 + 7	1 + 8	5 + 7	2 + 4	9 + 3
3 + 7	4 + 8	5 + 2	7 + 4	9 + 5
3 + 8	6 + 1	6 + 4	3 + 9	5 + 4

Name ______________________________

Set 9: Sums of 12; Review Facts

1. Read the answers to someone.
2. Write the answers.
3. Ask someone to correct your paper. Corrected by ______________________

$8 + 3$	$5 + 7$	$3 + 6$	$9 + 3$	$6 + 7$
$6 + 9$	$6 + 1$	$4 + 8$	$3 + 7$	$5 + 4$
$9 + 5$	$2 + 8$	$7 + 5$	$9 + 2$	$6 + 4$
$7 + 4$	$3 + 9$	$2 + 4$	$6 + 6$	$3 + 8$
$8 + 4$	$1 + 8$	$4 + 7$	$5 + 2$	$8 + 7$

Name ______________________________

Date ______________________________

Guided Class Practice 49A

Saxon Math 2 (for use with Lesson 49)

1. Steven had 6 nickels. Michelle gave Steven 2 more nickels. How many nickels does Steven have now?

Number sentence ______________________________

Answer ______________________________

How much money is that? ____________

2. Measure each line segment using inches.

horizontal line segment ________"

oblique line segment ________"

vertical line segment ________"

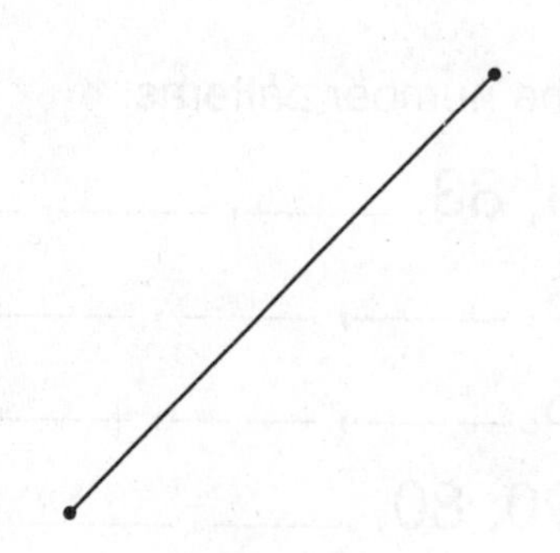

3. Write the numbers in order from least to greatest.

49	73	61

________ ________ ________
least greatest

4. Number the clockface.
Show five o'clock on the clockface.
It's five o'clock now.
Write the digital time one half hour from now.

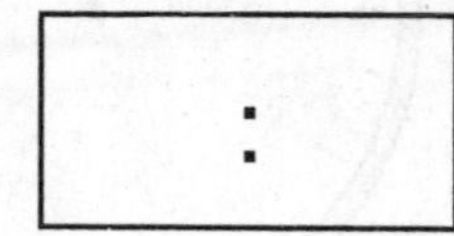

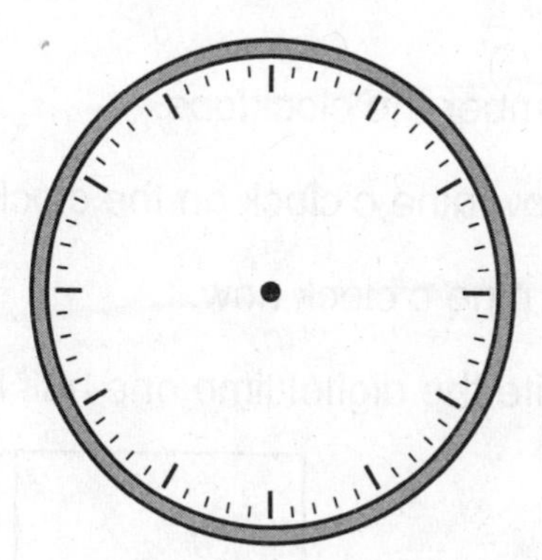

5. Find the sums.

26 + 10 = ______ 10 + 72 = ______ 50 + 10 = ______

6. Write the full date for 3/14/09. ______________________________

Name ____________________

Date ____________________

Homework 49B

Saxon Math 2 (for use with Lesson 49)

1. Marsha and Sandy are saving nickels. Marsha has 3 nickels and Sandy has 7 nickels. How many nickels do they have altogether?

   Number sentence ____________________

   Answer ____________________

   How much money is that? __________

2. Finish the number patterns.

   85, 84, 83, ____, ____, ____, ____, ____, ____, ____

   1, 3, 5, ____, ____, ____, ____, ____, ____, ____

   2, 4, 6, ____, ____, ____, ____, ____, ____, ____

   100, 90, 80, ____, ____, ____, ____, ____, ____, ____

3. Write the numbers in order from least to greatest.

52	38	21

   ____ least ____ ____ greatest

4. Number the clockface.

   Show nine o'clock on the clockface.

   It's nine o'clock now.

   Write the digital time one half hour from now.

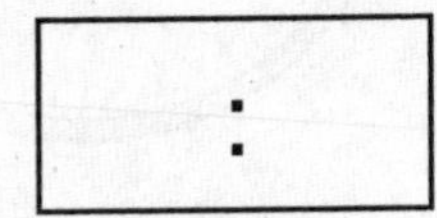

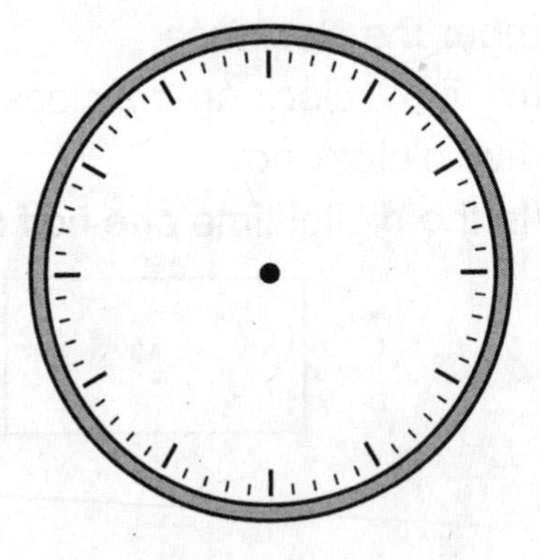

5. Find the sums.

   39 + 10 = ____ 10 + 19 = ____ 30 + 10 = ____

6. Write the full date for 7/9/07. ____________________

Name ______________ Score ________

Fact Assessment 9

Saxon *Math 2 (for use with* ***Lesson 50-1****)*

Set 9: Sums of 12; Review Facts

$8 + 3$	$5 + 7$	$1 + 5$	$7 + 4$	$3 + 6$
$4 + 8$	$6 + 9$	$2 + 8$	$9 + 3$	$5 + 2$
$4 + 7$	$1 + 8$	$7 + 5$	$9 + 2$	$6 + 4$
$3 + 9$	$5 + 4$	$3 + 0$	$8 + 4$	$6 + 1$
$3 + 6$	$2 + 4$	$6 + 6$	$3 + 7$	$9 + 5$

Name ____________________

Lesson Worksheet 50-1A

Saxon Math 2 (for use with Lesson 50-1)

A.

4 + ☐ = 13	5 + ☐ = 13	6 + ☐ = 13	7 + ☐ = 13	8 + ☐ = 13
9 + ☐ = 13	5 + ☐ = 14	6 + ☐ = 14	8 + ☐ = 14	9 + ☐ = 14

B.

6 + 7	8 + 6	4 + 9	7 + 7	8 + 5
8 + 6	9 + 4	6 + 8	5 + 9	7 + 6
5 + 8	9 + 5	6 + 7	7 + 7	4 + 9

Dear Parent,

Today we practiced the sum of 13 and 14 facts. Your child will be tested next week on the new addition facts listed below, along with previously introduced facts.

5 + 8	8 + 5	6 + 8	8 + 6

During the next four days, your child will practice these facts in class both orally and in writing. The back of each day's fact sheet will contain the fact practice homework.

Keep this paper at home to use for practice.

Name ____________________

Date ____________________

Problem-Solving Worksheet **50A**

***Saxon** Math 2 (for use with **Lesson 50-1**)*

Understand	Plan	Solve	Check

Make an Organized List

Riley has a cup of pennies and a cup of dimes. Show all the different ways he can make 45¢ using his dimes and pennies.

How many different ways can Riley make 45¢ using dimes and pennies? ________

Name ______________________

Performance Task Worksheet 50B

Saxon Math 2 (for use with Lesson 50-1)

Understand	Plan	Solve	Check

Jordan has a cup of pennies and a cup of dimes. Show all the different ways she can make 37¢ using her dimes and pennies.

How many different ways can Jordan make 37¢ using dimes and pennies? ________

Circle the problem-solving strategies you used to solve this problem.

Act It Out ***Use Logical Reasoning***

Draw a Picture ***Look for a Pattern***

Make an Organized List

Explain how you got your answer: ______________________________

____

____

____

Name ___________________________

Date ___________________________

1. Christa had 8 markers and 3 pencils. She gave 2 markers to her friend. How many markers does she have now?

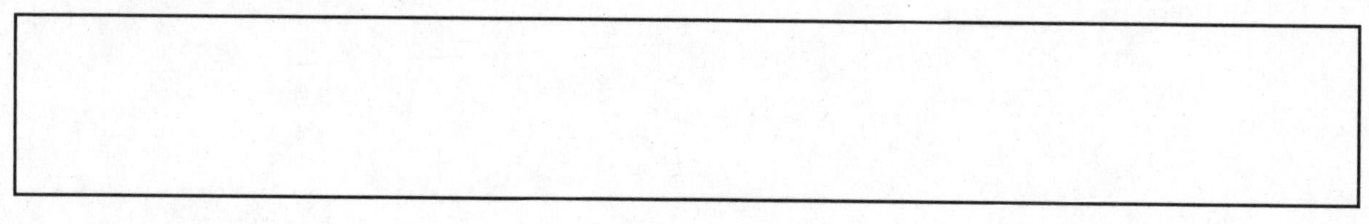

Number sentence ___________________________

Answer ___________________________

2. Use a **yellow** crayon to trace the **oblique** line segment.
Use a **red** crayon to trace the **horizontal** line segment.
Use a **blue** crayon to trace the **vertical** line segment.

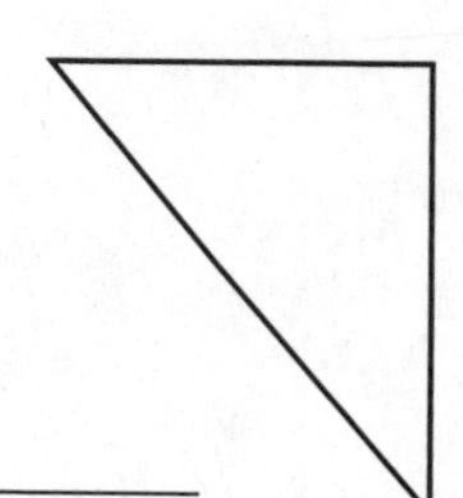

3. I have 4 pennies and 8 dimes. How much money is this? ________

Make 37¢ using the fewest dimes and pennies.

________ dimes ________ pennies

4. How many shoes are there? ________ shoes

Circle pairs of shoes.

How many pairs of shoes are there? ________ pairs

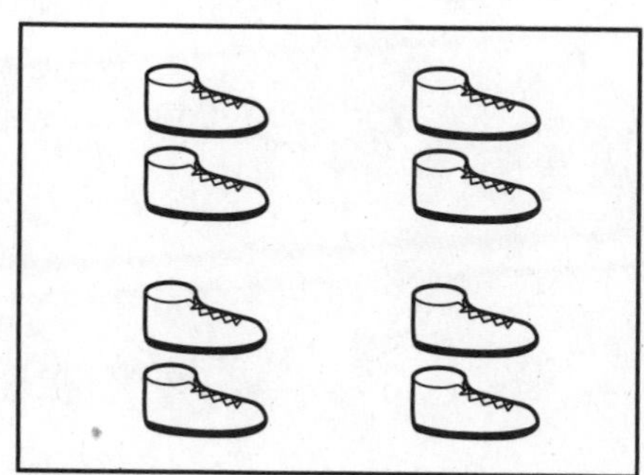

5. Find the sums.

$3 + 7 =$ ________ $5 + 6 =$ ________

$5 + 2 =$ ________ $4 + 3 =$ ________

6. Fill in the missing numbers on this piece of a hundred number chart.

23				
			36	
	44			

Name ______________________________

Class Fact Practice 51A

***Saxon** Math 2 (for use with **Lesson 51**)*

Set 10: Sums of 13 and 14; Review Facts

$7 + 3$	$8 + 5$	$5 + 6$	$7 + 8$	$5 + 9$
$4 + 8$	$7 + 6$	$0 + 6$	$8 + 6$	$7 + 5$
$5 + 8$	$7 + 4$	$4 + 9$	$3 + 8$	$7 + 7$
$4 + 7$	$9 + 5$	$4 + 6$	$6 + 7$	$8 + 4$
$3 + 9$	$6 + 8$	$8 + 3$	$9 + 4$	$5 + 7$

Name ______________________________

Fact Homework 51B

Saxon Math 2 (for use with Lesson 51)

Set 10: Sums of 13 and 14; Review Facts

1. Read the answers to someone.
2. Write the answers.
3. Ask someone to correct your paper. Corrected by ____________________

6 + 7	4 + 8	7 + 7	7 + 4	9 + 5
7 + 5	8 + 3	5 + 8	7 + 3	7 + 8
3 + 8	9 + 4	5 + 7	8 + 6	0 + 6
8 + 5	4 + 7	5 + 9	7 + 6	4 + 6
3 + 9	6 + 8	8 + 4	5 + 6	4 + 9

Name ______________________________

A.	___ dimes	___ nickels	___ pennies	¢
B.	___ dimes	___ nickels	___ pennies	¢
C.	___ dimes	___ nickels	___ pennies	¢
D.	___ dimes	___ nickels	___ pennies	¢

Name ____________________

Date ____________________

Guided Class Practice 51A

***Saxon** Math 2 (for use with **Lesson 51**)*

1. Amy's dog had 8 puppies. She gave five puppies to her friends and one to her grandmother. How many puppies does Amy have now?

Number sentence ____________________

Answer ____________________

2. Measure this line segment using inches.

______ "

3. Write the numbers in order from least to greatest.

65	54	63

______ least ______ ______ greatest

4. How much money is this? ____________

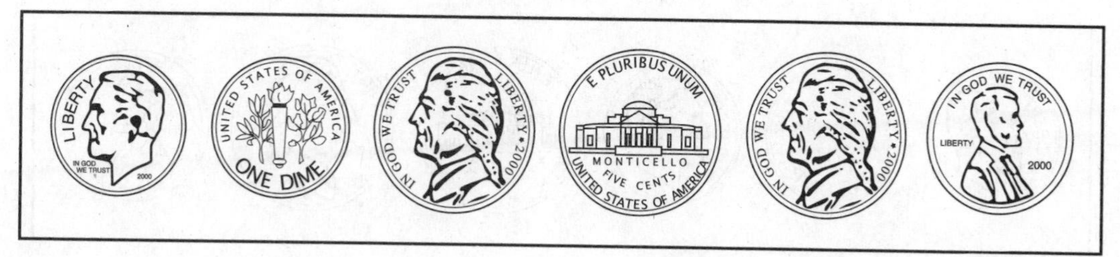

5. Show the times on the clocks.

Name

Date

Homework 51B

***Saxon** Math 2 (for use with **Lesson 51**)*

1. Beth's cat had 5 kittens. She gave 2 kittens to her teacher and one to her aunt. How many kittens does Beth have now?

Number sentence ____

Answer ____

2. This is a six-inch line segment. Find something at home that is 6" long.

What did you find? ____

3. Write the numbers in order from least to greatest.

86 81 74

____ ____ ____
least greatest

4. How much money is this? ____

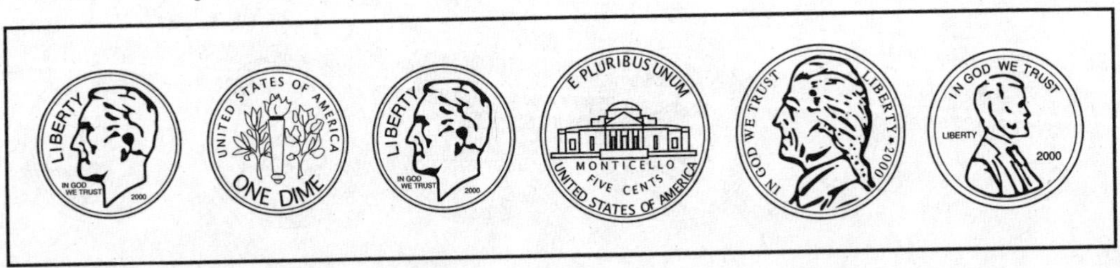

5. Show the times on the clocks.

Name ______________________________

Class Fact Practice 52A

***Saxon** Math 2 (for use with **Lesson 52**)*

Set 10: Sums of 13 and 14; Review Facts

$4 + 9$	$5 + 6$	$8 + 4$	$6 + 8$	$3 + 9$
$4 + 6$	$7 + 6$	$5 + 9$	$4 + 7$	$8 + 5$
$0 + 6$	$8 + 6$	$5 + 7$	$9 + 4$	$3 + 8$
$7 + 8$	$7 + 3$	$5 + 8$	$8 + 3$	$7 + 5$
$9 + 5$	$7 + 4$	$7 + 7$	$4 + 8$	$6 + 7$

Name ________________________________

Fact Homework 52B

***Saxon** Math 2 (for use with **Lesson 52**)*

Set 10: Sums of 13 and 14; Review Facts

1. Read the answers to someone.
2. Write the answers.
3. Ask someone to correct your paper. Corrected by ____________________

$9 + 4$	$7 + 5$	$4 + 6$	$8 + 5$	$5 + 6$
$8 + 4$	$7 + 6$	$8 + 3$	$6 + 8$	$4 + 7$
$7 + 8$	$9 + 5$	$7 + 3$	$3 + 9$	$7 + 7$
$5 + 8$	$7 + 4$	$0 + 6$	$4 + 9$	$3 + 8$
$5 + 7$	$8 + 6$	$4 + 8$	$5 + 9$	$6 + 7$

Name ______________________________

Date ______________________________

1. Nathan has 8 pencils in his desk. Keith has 7 pencils in his desk. How many pencils do they have altogether?

   Number sentence ______________________________

   Answer ______________________________

2. Measure each side of this square.

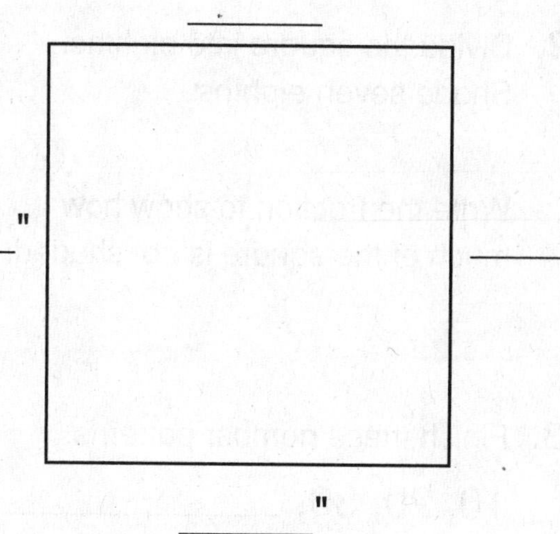

3. Divide the square into fourths. Shade three fourths.

   Write the fraction to show how much of the square is not shaded. ________

4. How much money is this? ____________

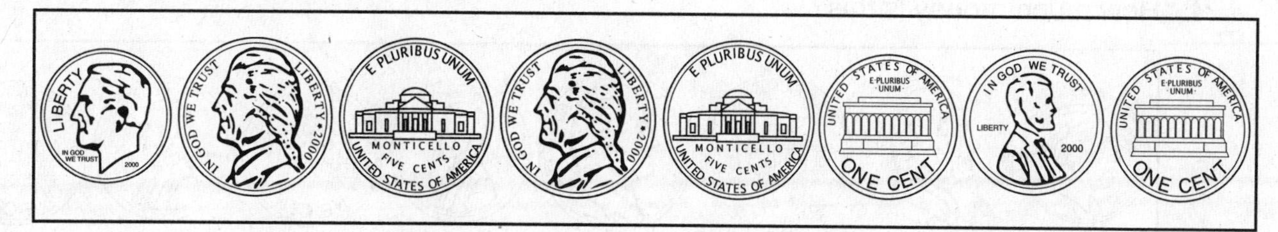

5. Draw a line of symmetry for each shape.

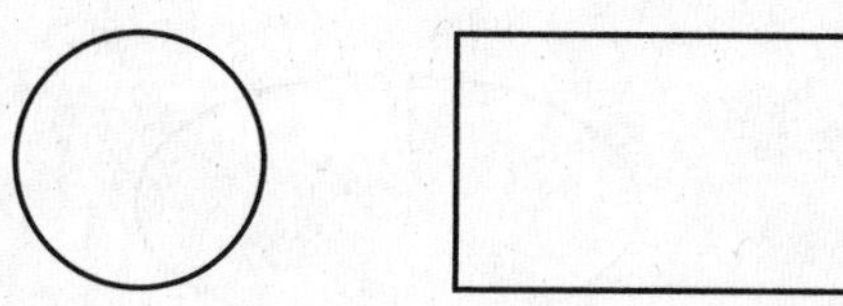

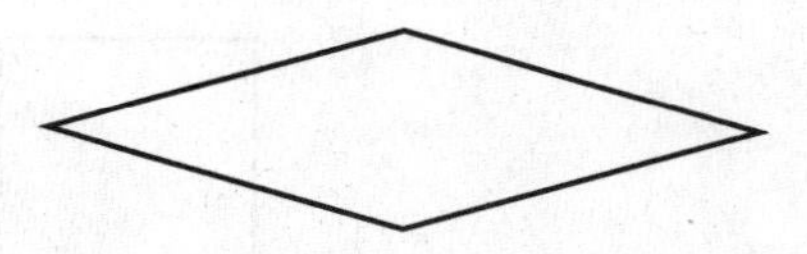

6. Write the numbers in order from least to greatest.

52	21	57

________ ________ ________
least greatest

Name ______________________________

Date ______________________________

1. Beth had twelve books. She gave one to her brother. How many books does she have now?

   Number sentence ______________________________

   Answer ______________________

2. Divide the square into eighths. Shade seven eighths.

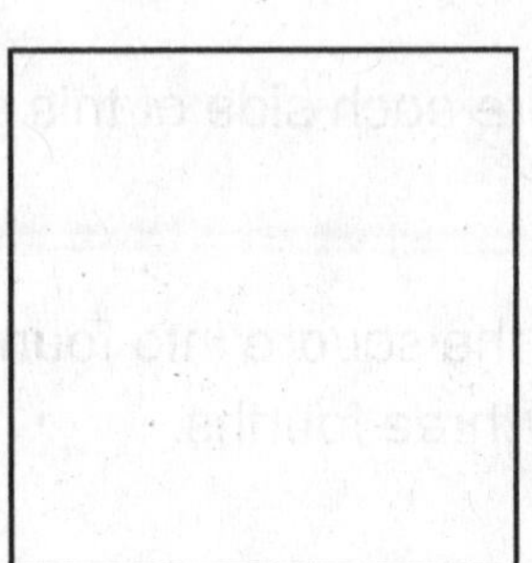

   Write the fraction to show how much of the square is not shaded. ______

3. Finish these number patterns.

   10, 20, 30, ______, ______, ______, ______, ______, ______, ______

   50, 55, 60, 65, ______, ______, ______, ______, ______, ______, ______

4. How much money is this? __________

5. Draw a line of symmetry for each shape.

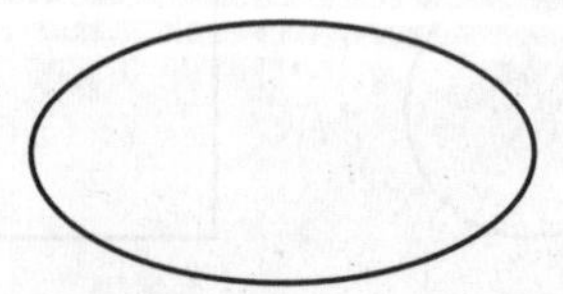

6. Write the numbers in order from least to greatest.

75	46	43

______ least ______ ______ greatest

Name ____________________

Set 10: Sums of 13 and 14; Review Facts

A.

$4 + 9 =$ ____ $8 + 4 =$ ____ $6 + 7 =$ ____

$7 + 4 =$ ____ $5 + 9 =$ ____ $4 + 6 =$ ____

$3 + 9 =$ ____ $5 + 6 =$ ____ $6 + 8 =$ ____

$5 + 7 =$ ____ $9 + 4 =$ ____ $7 + 3 =$ ____

$5 + 8 =$ ____ $4 + 7 =$ ____ $9 + 5 =$ ____

$4 + 8 =$ ____ $8 + 6 =$ ____ $7 + 5 =$ ____

$7 + 6 =$ ____ $3 + 8 =$ ____ $8 + 5 =$ ____

B.

$8 + \square = 14$ $7 + \square = 14$ $5 + \square = 14$

$\square + 6 = 13$ $\square + 4 = 13$ $\square + 8 = 13$

Name ______________________________

Saxon Math 2 (for use with **Lesson 53**)

Set 10: Sums of 13 and 14; Review Facts

A. 1. Read the answers to someone.
2. Write the answers.
3. Ask someone to correct your paper. Corrected by ______________________

8 + 5 = ____	3 + 8 = ____	7 + 6 = ____
9 + 5 = ____	4 + 7 = ____	5 + 8 = ____
7 + 5 = ____	8 + 6 = ____	4 + 8 = ____
7 + 3 = ____	9 + 4 = ____	5 + 7 = ____
6 + 8 = ____	5 + 6 = ____	3 + 9 = ____
4 + 6 = ____	5 + 9 = ____	7 + 4 = ____
6 + 7 = ____	8 + 4 = ____	4 + 9 = ____

B. Fill in the missing numbers.

6 + ☐ = 14	7 + ☐ = 14	9 + ☐ = 14
☐ + 9 = 13	☐ + 7 = 13	☐ + 5 = 13

Name ____________________

1.

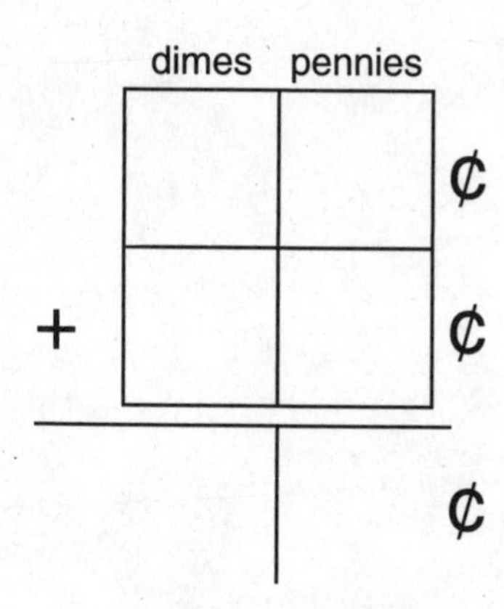

2.

dimes pennies

¢

\+ ¢

¢

3.

dimes pennies

¢

\+ ¢

¢

4.

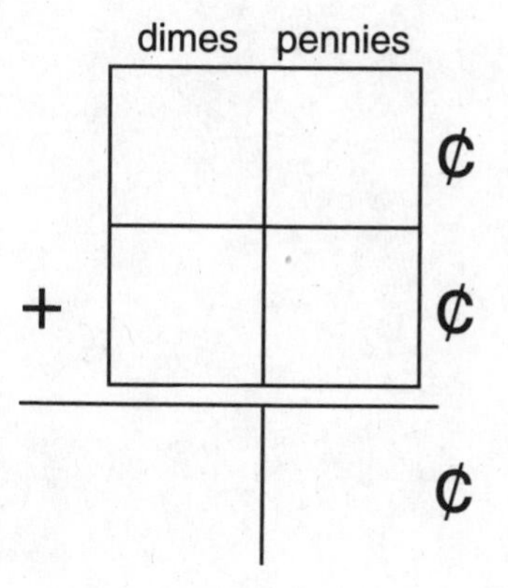

5.

dimes pennies

¢

\+ ¢

¢

6.

dimes pennies

¢

\+ ¢

¢

7.

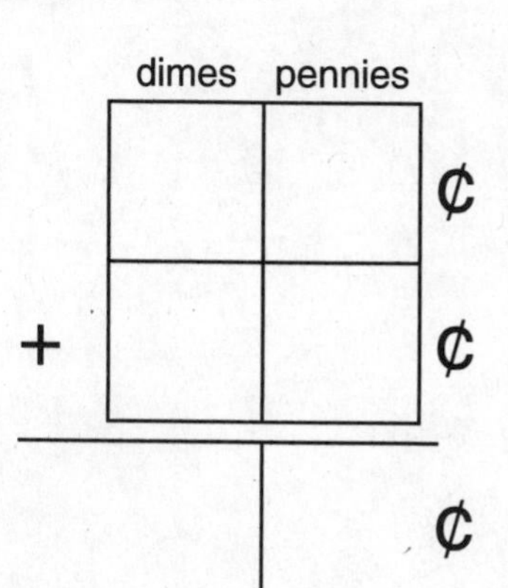

8.

dimes pennies

¢

\+ ¢

¢

9.

dimes pennies

¢

\+ ¢

¢

Name ______________________________

Date ______________________________

1. Mrs. Hannan's classroom has 32 pairs of right-handed scissors and 10 pairs of left-handed scissors. How many pairs of scissors are there in all?

Number sentence ______________________________

Answer ______________________________

2. Draw a line of symmetry for each letter.

M O D

3. How much money is this? __________

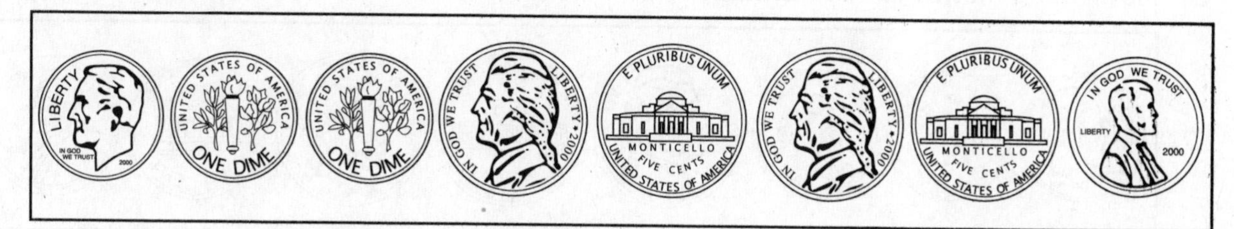

4. One of these names is my mother's name.
Use the clues to find my mother's name.
Cross out the names that cannot be my mother's name.

It does not have exactly six letters.
It does not begin with a vowel.
It is not first.
It is not fifth.
Circle my mother's name.

Mary	Bernice	Louise	Anna	Cora

5. Find the sums.

78 + 10 = ______ 35 + 10 = ______ 23 + 10 = ______

6. I have **63¢.** How many dimes and pennies is that?

I have **24¢.** How many dimes and pennies is that?

How many dimes and pennies is that altogether?

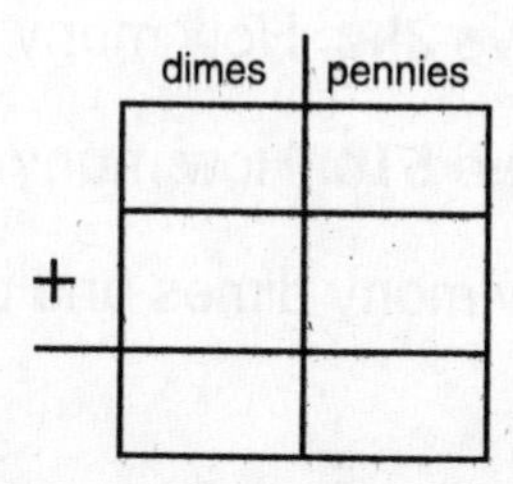

Name ______________________

Date ______________________

Saxon Math 2 (for use with Lesson 53)

1. Ryan had 64 baseball cards. His brother gave him 10 more cards. How many cards does he have now?

Number sentence ______________________

Answer ______________________

2. Draw a line of symmetry for each letter.

H V X

3. How much money is this? __________

4. One of these names is my brother's name.
Use the clues to find my brother's name.
Cross out the names that cannot be my brother's name.

It does not begin with a vowel.
It is not second.
It does not have exactly three letters.
Circle my brother's name.

Arthur	Paul	David	Sam

5. Find the sums.

65 + 10 = ______ 57 + 10 = ______ 37 + 10 = ______

6. I have **34¢.** How many dimes and pennies is that?

I have **51¢.** How many dimes and pennies is that?

How many dimes and pennies is that altogether?

	dimes	pennies
+		

Name ______________________________

Class Fact Practice 54A

***Saxon** Math 2 (for use with **Lesson 54**)*

Set 10: Sums of 13 and 14; Review Facts

6 + 7	5 + 9	4 + 8	8 + 6	5 + 7
3 + 8	4 + 9	0 + 6	7 + 4	5 + 8
7 + 7	3 + 9	7 + 3	9 + 5	7 + 8
4 + 7	6 + 8	8 + 3	7 + 6	8 + 4
5 + 6	8 + 5	4 + 6	7 + 5	9 + 4

Name ______________________________

Fact Homework 54B

***Saxon** Math 2 (for use with **Lesson 54**)*

Set 10: Sums of 13 and 14; Review Facts

1. Read the answers to someone.
2. Write the answers.
3. Ask someone to correct your paper. Corrected by ____________________

$5 + 7$	$9 + 4$	$8 + 3$	$6 + 8$	$3 + 9$
$8 + 4$	$6 + 7$	$4 + 6$	$9 + 5$	$4 + 7$
$7 + 7$	$3 + 8$	$4 + 9$	$7 + 4$	$5 + 8$
$7 + 5$	$8 + 6$	$0 + 6$	$7 + 6$	$4 + 8$
$5 + 9$	$7 + 8$	$5 + 6$	$8 + 5$	$7 + 3$

Name ______________________

Date ______________________

1. Weston had 60¢. His sister gave him ten cents. How much money does he have now?

What kind of problem is this? ______________________

Number sentence ______________________

Answer ______________________

2. Add. 54¢ + 31¢ = ________ 27¢ + 52¢ = ________

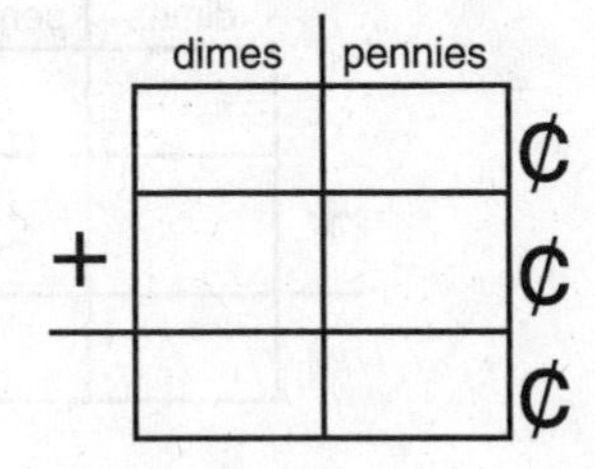

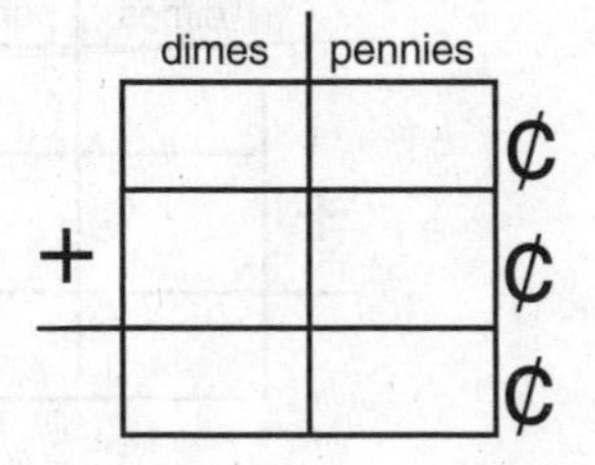

3. How much money is this? ________

4. Find the sums.

55 + 10 = ________ 27 + 10 = ________ 40 + 10 = ________

5. Measure this line segment using inches.

________"

6. Fill in the missing numbers in this fact family.

☐ + 8 = 13 8 + ☐ = 13

13 − ☐ = 5 ☐ − 5 = 8

Name ______________________

Date ______________________

Homework 54B

***Saxon** Math 2 (for use with **Lesson 54**)*

1. Stephen had 50¢. He gave his sister ten cents. How much money does he have now?

What kind of problem is this? ______________________

Number sentence ______________________

Answer ______________________

2. Add. 25¢ + 34¢ = ________ 83¢ + 15¢ = ________

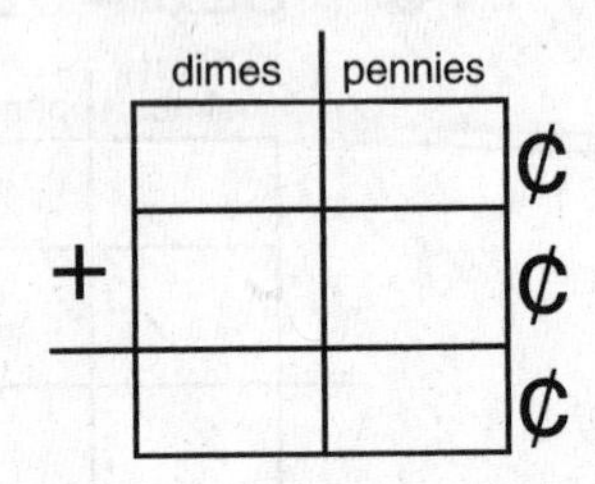

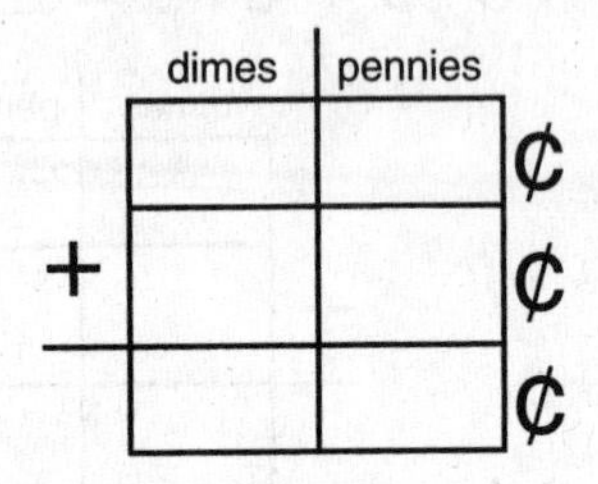

3. Ask someone in your family to let you count the change in his or her pocket or purse. (Don't count the quarters.)

How many dimes are there? ________ How much money is that? ________

How many nickels are there? ________ How much money is that? ________

How many pennies are there? ________ How much money is that? ________

How much money is this altogether? ________

4. Find the sums.

24 + 10 = ________ 46 + 10 = ________ 61 + 10 = ________

5. Finish these number patterns.

5, 10, 15, ____, ____, ____, ____, ____, ____, ____

1, 3, 5, ____, ____, ____, ____, ____, ____, ____

6. Fill in the missing numbers in this fact family.

☐ + 8 = 14 ☐ + 6 = 14

☐ − 6 = 8 ☐ − 8 = 6

Name ______________________ Score ______

Fact Assessment 10

***Saxon** Math 2 (for use with **Lesson 55-1**)*

Set 10: Sums of 13 and 14; Review Facts

$3 + 9$	$6 + 8$	$7 + 7$	$9 + 4$	$5 + 7$
$7 + 3$	$8 + 5$	$5 + 6$	$7 + 8$	$5 + 9$
$4 + 8$	$7 + 6$	$0 + 6$	$8 + 6$	$7 + 1$
$5 + 8$	$8 + 2$	$4 + 9$	$3 + 8$	$7 + 7$
$4 + 7$	$9 + 5$	$4 + 6$	$6 + 7$	$8 + 4$

Name ____________________

A.

$6 + \square = 15$	$7 + \square = 15$	$8 + \square = 15$	$9 + \square = 15$	$7 + \square = 16$
$8 + \square = 16$	$9 + \square = 16$	$8 + \square = 17$	$9 + \square = 17$	$9 + \square = 18$

B.

$8 + 7$	$9 + 9$	$7 + 9$	$8 + 9$	$9 + 6$
$8 + 8$	$9 + 8$	$6 + 9$	$9 + 7$	$7 + 8$
$9 + 6$	$8 + 9$	$7 + 9$	$8 + 7$	$9 + 8$

Name ______________________________

Date ______________________________

Guided Class Practice 55A

Saxon *Math 2 (for use with* ***Lesson 55-1****)*

1. Gilbert's mom made a pizza and cut it into 8 equal pieces.
Show how she cut the pizza.
Gilbert ate half of the pizza.
Color the part of the pizza Gilbert ate.

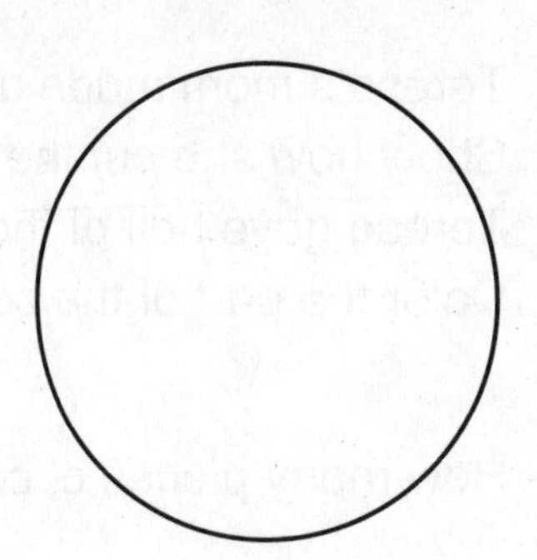

How many pieces did Gilbert eat? ________

2. Fill in the missing numbers on this piece of a hundred number chart.

			37		
		46			

3. Color three fourths green.
Color one half red.
Color three eighths yellow.

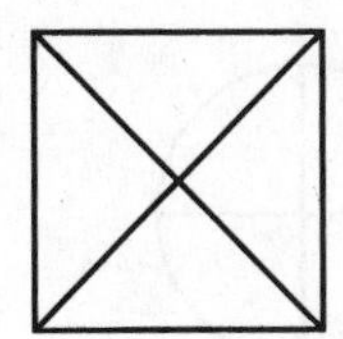
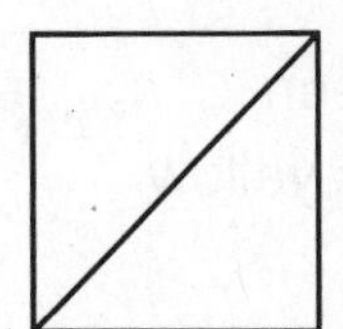
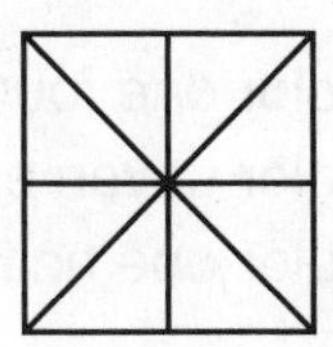

4. What time is shown on the clock?

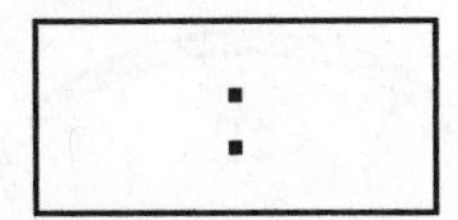

What time was it one half hour ago?

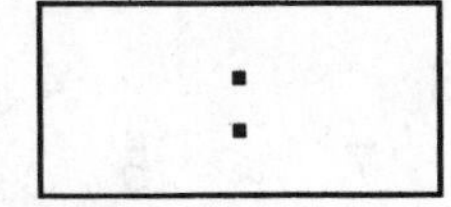

5. Find the sums.

	dimes	pennies	
	7	3	¢
+	1	4	¢
			¢

	dimes	pennies	
	4	6	¢
+	5	2	¢
			¢

Name ____________________________

Date ____________________________

Homework 55B

***Saxon** Math 2 (for use with **Lesson 55-1**)*

1. Teresa's mom made a cake and cut it into 8 equal pieces.
Show how she cut the cake.
Teresa gave half of the cake to her grandfather.
Color the part of the cake Teresa gave her grandfather.

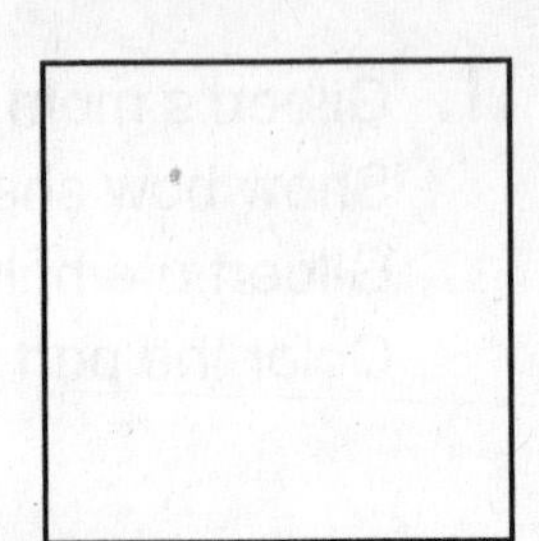

How many pieces of cake did she give her grandfather? ______

2. Fill in the missing numbers on this piece of a hundred number chart.

		74			
82					

3. Color one fourth green.
Color seven eighths yellow.
Color one half red.

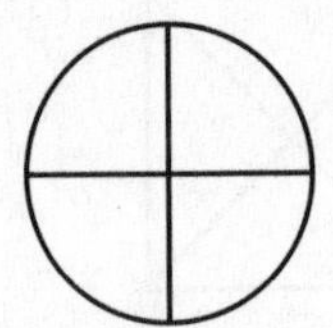

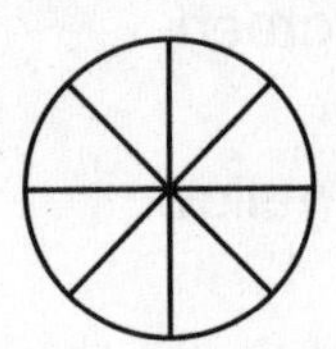

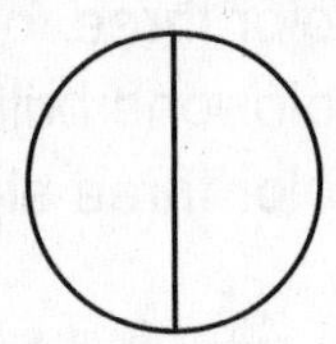

4. What time do you usually go to bed at night?
Show the time on the clocks.

:

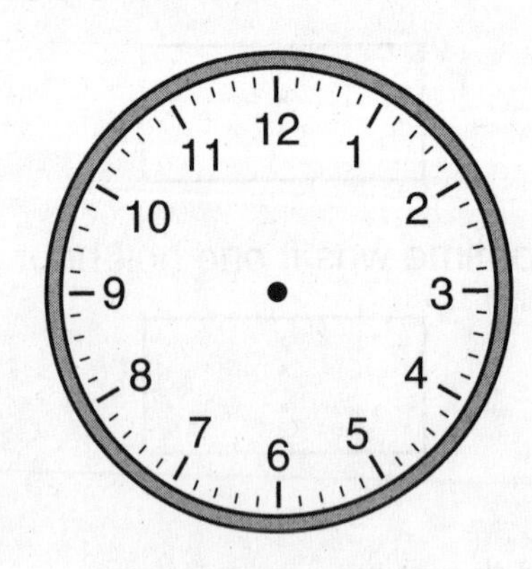

5. Find the sums.

	dimes	pennies	
	2	3	¢
+	4	3	¢
			¢

	dimes	pennies	
	5	5	¢
+	3	4	¢
			¢

Name ______________________________

Date ______________________________

Written Assessment 10

Saxon Math 2 (for use with Lesson 55-2)

1. Courtney had a box of 64 crayons. Sharon gave her 10 more crayons. How many crayons does Courtney have now?

Number sentence ______________________________

Answer ______________________________

2. One of these is my favorite color. Use the clues to find my favorite color. Cross out the colors that cannot be my favorite color.

It is not in the middle.
It does not have exactly four letters.
It is not fifth.
It is not second.
Circle my favorite color.

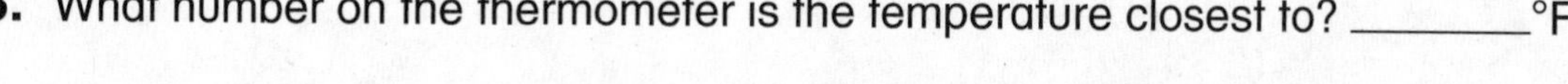

3. What number on the thermometer is the temperature closest to? ______ °F

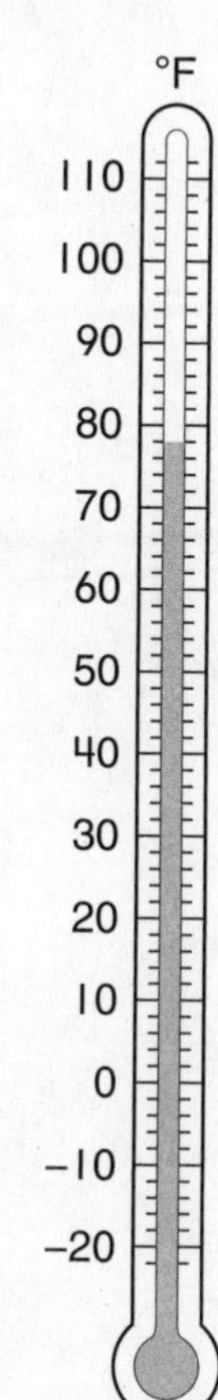

4. Measure each line segment using inches.

______ "

______ "

5. Finish these number patterns.

2, 4, 6, ____, ____, ____, ____, ____, ____, ____

1, 3, 5, ____, ____, ____, ____, ____, ____, ____

6. Find each answer.

46 + 10 = ______ 10 + 37 = ______ 74 + 10 = ______

Name ______________________

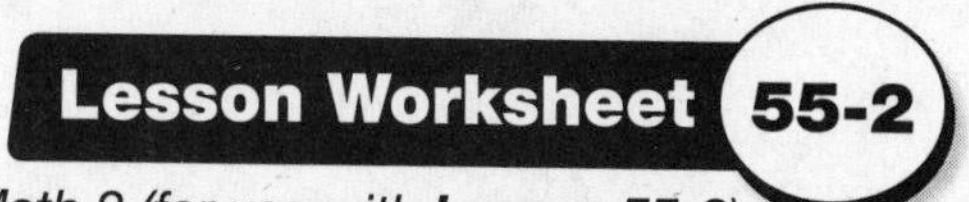

Saxon *Math 2 (for use with* ***Lesson 55-2****)*

What we measured	________'s feet	________'s feet	Ruler feet
1.			
2.			
3.			
4.			

Name ______________________

Class Fact Practice 56A

***Saxon** Math 2 (for use with **Lesson 56**)*

Set 11: Sums of 15, 16, 17, and 18; Review Facts

$9 + 4$	$7 + 5$	$8 + 7$	$6 + 3$	$7 + 9$
$8 + 5$	$9 + 8$	$4 + 5$	$7 + 8$	$5 + 3$
$9 + 6$	$3 + 8$	$5 + 7$	$9 + 9$	$8 + 6$
$4 + 3$	$9 + 7$	$4 + 8$	$5 + 9$	$8 + 8$
$6 + 9$	$3 + 7$	$5 + 8$	$8 + 9$	$6 + 8$

Name ______________________________

Fact Homework 56B

Saxon *Math 2 (for use with* ***Lesson 56****)*

Set 11: Sums of 15, 16, 17, and 18; Review Facts

1. Read the answers to someone.
2. Write the answers.
3. Ask someone to correct your paper. Corrected by ____________________

8 + 8	5 + 9	4 + 8	9 + 7	4 + 3
6 + 8	8 + 9	5 + 8	3 + 7	6 + 9
7 + 9	6 + 3	8 + 7	7 + 5	9 + 4
5 + 3	7 + 8	4 + 5	9 + 8	8 + 5
8 + 6	9 + 9	5 + 7	3 + 8	9 + 6

Name ______________________________

1. Write the missing numbers below the number line.
Put a point at 7 on the number line.
Write the letter *A* above the point.
Put a point at 4 on the number line.
Write the letter *B* above the point.

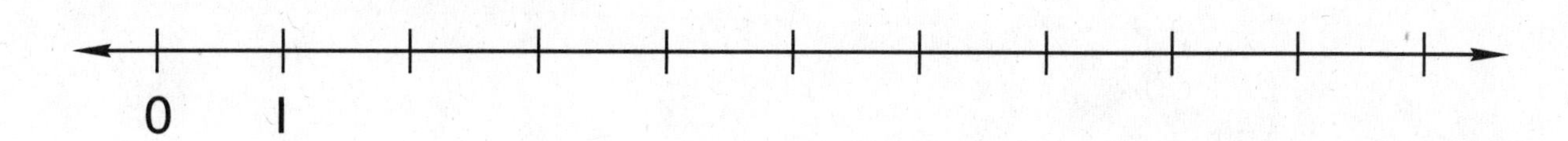

2. Write the missing numbers below the number line.
Put a point at 5 on the number line.
Write the letter *C* above the point.
Put a point at 9 on the number line.
Write the letter *D* above the point.

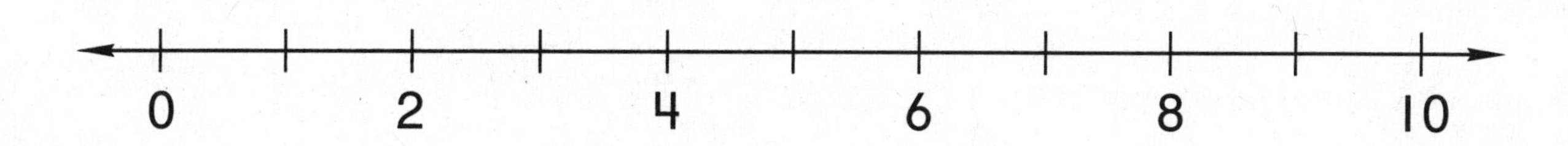

3. Write the missing numbers below the number line.

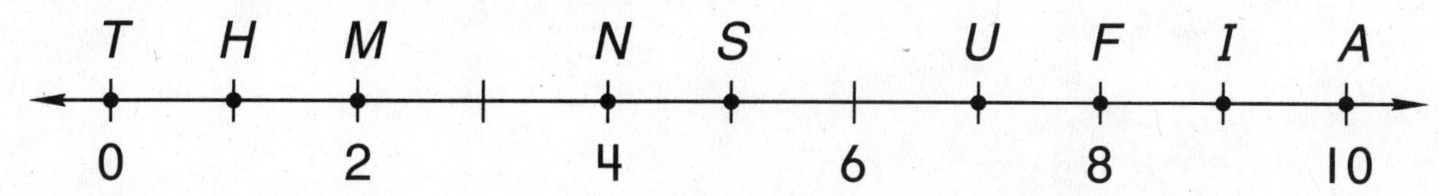

Which letters name the points on the number line?

___	___	___	___		___	___		___	___	___
2	10	0	1		9	5		8	7	4

Name ______________________

Date ______________________

Guided Class Practice 56A

***Saxon** Math 2 (for use with **Lesson 56**)*

1. There are 10 boys and 8 girls in Room 7. Two girls went home sick. How many girls are in Room 7 now?

Number sentence ______________________

Answer ______________________

2. Write the numbers below the number line.
Put a point at **3.** Label it **N.**
Put a point at **5.** Label it **D.**

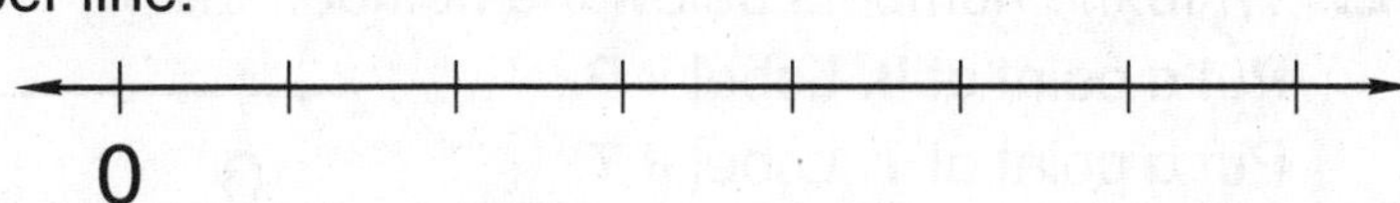

0

3. Draw a 4-inch line segment.

•

4. Draw 4 pairs of mittens. Circle the pairs.

How many fingers will be in the mittens? ______ fingers
Show that number using tally marks.

5. Find the sums.

10 + 25 = ______

58 + 10 = ______

	dimes	pennies	
	2	1	¢
+	2	3	¢
			¢

	dimes	pennies	
	5	4	¢
+	3	5	¢
			¢

Name ______________________

Date ______________________

Homework 56B

Saxon *Math 2 (for use with* ***Lesson 56****)*

1. Ivan's dog ate 9 dog biscuits on Sunday. On Monday he ate 10 more. How many dog biscuits did he eat altogether?

Number sentence ______________________

Answer ______________________

2. Write the numbers below the number line.
Put a point at **4.** Label it **P.**
Put a point at **1.** Label it **T.**

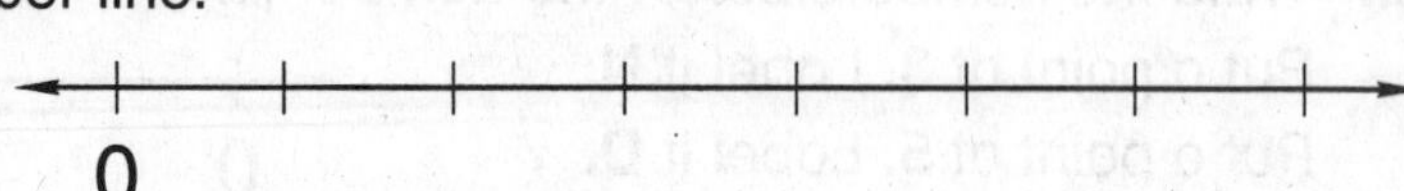

3. Which number on the thermometer is the temperature closest to? ______ °F

°F
110
100
90
80
70
60
50
40
30
20
10
0
−10
−20

4. Draw 3 pairs of socks. Circle the pairs.

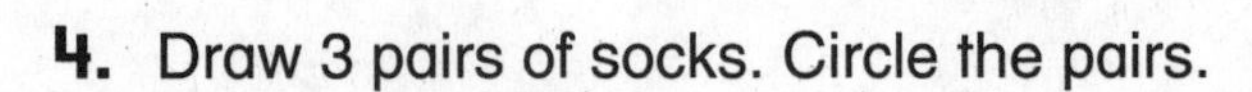

How many toes will be in the socks? ______ toes
Show that number using tally marks.

5. Find the sums.

10 + 35 = ______

63 + 10 = ______

	dimes	pennies	
	3	6	¢
+	4	2	¢
			¢

	dimes	pennies	
	1	4	¢
+	7	4	¢
			¢

Name ______________________

Class Fact Practice **57A**

***Saxon** Math 2 (for use with **Lesson** 57)*

Set 11: Sums of 15, 16, 17, and 18; Review Facts

5 + 7	9 + 6	3 + 7	5 + 8	7 + 9
4 + 8	8 + 7	4 + 3	9 + 8	8 + 6
3 + 8	9 + 4	8 + 8	4 + 5	6 + 9
9 + 9	5 + 9	7 + 8	6 + 3	9 + 7
8 + 5	5 + 3	8 + 9	7 + 5	6 + 8

Name ______________________

Fact Homework 57B

***Saxon** Math 2 (for use with **Lesson 57**)*

Set 11: Sums of 15, 16, 17, and 18: Review Facts

1. Read the answers to someone.
2. Write the answers.
3. Ask someone to correct your paper. Corrected by ______________________

$$\begin{array}{r}8\\+\ 8\\\hline\end{array}\quad\begin{array}{r}5\\+\ 9\\\hline\end{array}\quad\begin{array}{r}4\\+\ 5\\\hline\end{array}\quad\begin{array}{r}6\\+\ 9\\\hline\end{array}\quad\begin{array}{r}7\\+\ 5\\\hline\end{array}$$

$$\begin{array}{r}8\\+\ 6\\\hline\end{array}\quad\begin{array}{r}7\\+\ 9\\\hline\end{array}\quad\begin{array}{r}8\\+\ 5\\\hline\end{array}\quad\begin{array}{r}5\\+\ 3\\\hline\end{array}\quad\begin{array}{r}9\\+\ 9\\\hline\end{array}$$

$$\begin{array}{r}8\\+\ 7\\\hline\end{array}\quad\begin{array}{r}4\\+\ 3\\\hline\end{array}\quad\begin{array}{r}5\\+\ 8\\\hline\end{array}\quad\begin{array}{r}9\\+\ 7\\\hline\end{array}\quad\begin{array}{r}4\\+\ 8\\\hline\end{array}$$

$$\begin{array}{r}9\\+\ 8\\\hline\end{array}\quad\begin{array}{r}3\\+\ 7\\\hline\end{array}\quad\begin{array}{r}7\\+\ 8\\\hline\end{array}\quad\begin{array}{r}9\\+\ 4\\\hline\end{array}\quad\begin{array}{r}6\\+\ 8\\\hline\end{array}$$

$$\begin{array}{r}3\\+\ 8\\\hline\end{array}\quad\begin{array}{r}9\\+\ 6\\\hline\end{array}\quad\begin{array}{r}6\\+\ 3\\\hline\end{array}\quad\begin{array}{r}8\\+\ 9\\\hline\end{array}\quad\begin{array}{r}5\\+\ 7\\\hline\end{array}$$

Name ______________________

Date ______________________

Guided Class Practice 57A

***Saxon** Math 2 (for use with **Lesson 57**)*

1. Selina has 6 dimes. Rhonda has 9 nickels. How much money does each girl have?

Selina __________ Rhonda __________

Who has more money? __________

2. Beth is putting some numbers in order from least to greatest. Which number in the box should she put between **43** and **67?**

41	58	35	69

43 (least) ______ 67 (greaest)

3. Put a dot inside each angle. Count the number of angles in each polygon.

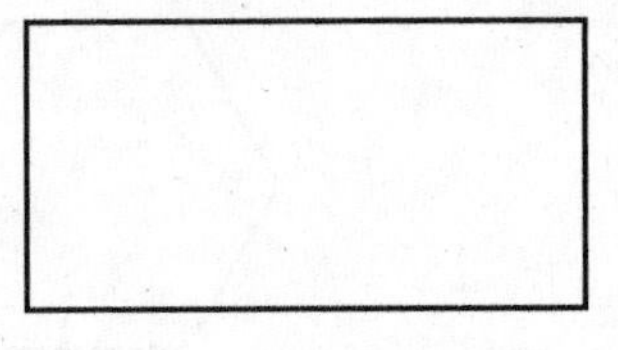

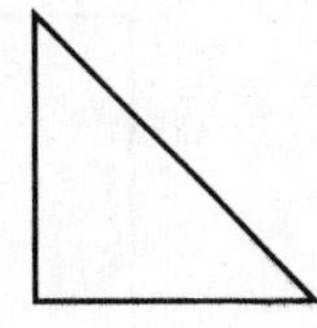

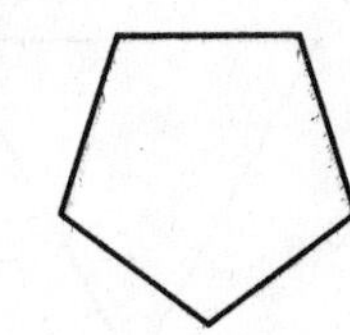

______ angles ______ angles ______ angles

4. Write the numbers below the number line.
Put a point at **4.** Label it **P.**
Put a point at **1.** Label it **T.**

Where is point **Z?** ______

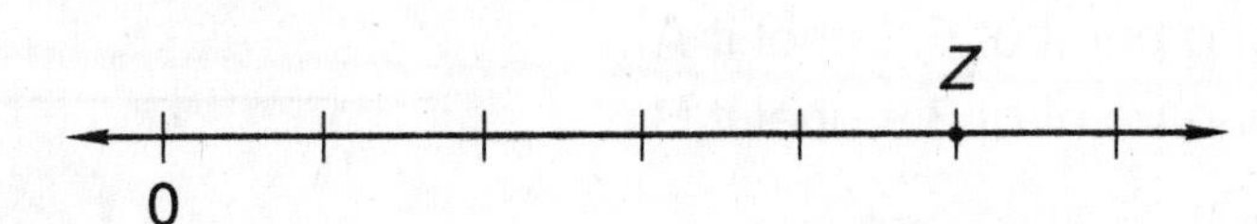

5. Measure each line segment using inches.

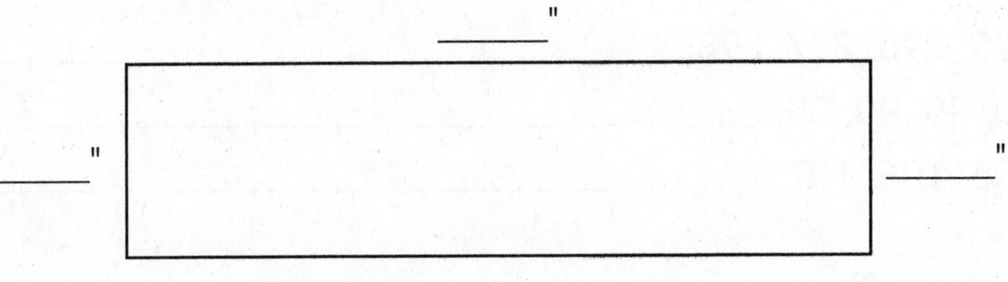

Name ______________________________

Date ______________________________

Homework 57B

***Saxon** Math 2 (for use with **Lesson 57**)*

1. Albert has 7 nickels. Luis has 3 dimes. How much money does each boy have?

Albert ____________ Luis ____________

Who has more money? ____________

2. Nora is putting some numbers in order from least to greatest. Which number in the box should she put between **57** and **82?**

48	84	53	71

57 (least) ______ 82 (greatest)

3. Put a dot inside each angle. Count the number of angles in each polygon.

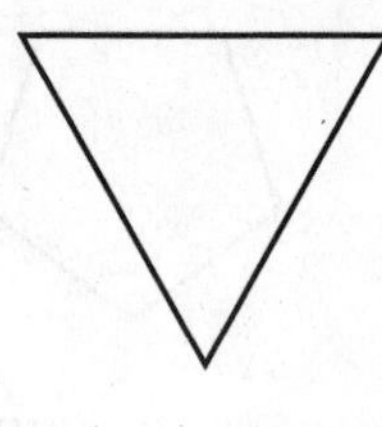

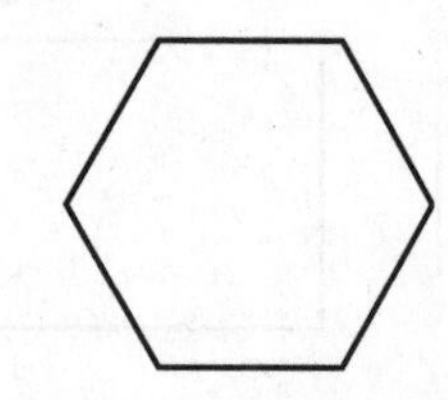

______ angles ______ angles ______ angles

4. Write the numbers below the number line.
Put a point at **5.** Label it **A.**
Put a point at **3.** Label it **B.**

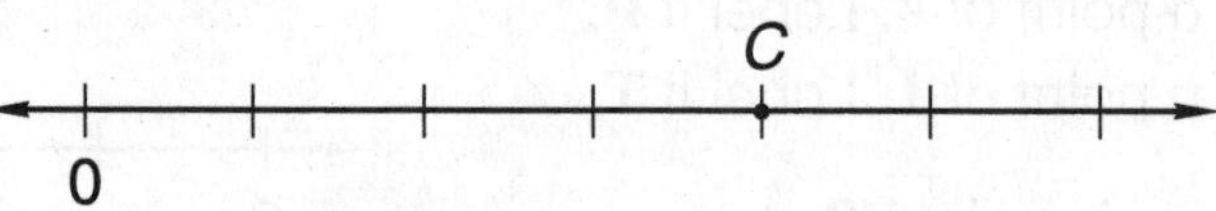

Where is point **C?** ______

5. Finish these number patterns.

295, 296, 297, 298, ______, ______, ______, ______, ______, ______

50, 45, 40, 35, ______, ______, ______, ______, ______, ______

105, 106, 107, ______, ______, ______, ______, ______, ______, ______

Name ______________________

Set 11: Sums of 15, 16, 17, and 18; Review Facts

A.

$5 + 7 =$ ____	$8 + 9 =$ ____	$6 + 8 =$ ____
$9 + 6 =$ ____	$6 + 3 =$ ____	$7 + 8 =$ ____
$9 + 4 =$ ____	$3 + 8 =$ ____	$7 + 9 =$ ____
$3 + 7 =$ ____	$8 + 7 =$ ____	$5 + 8 =$ ____
$9 + 8 =$ ____	$4 + 8 =$ ____	$5 + 9 =$ ____
$5 + 3 =$ ____	$9 + 7 =$ ____	$8 + 5 =$ ____
$8 + 6 =$ ____	$7 + 5 =$ ____	$6 + 9 =$ ____

B.

$9 + \square = 15$	$7 + \square = 15$	$6 + \square = 15$
$\square + 9 = 16$	$\square + 8 = 16$	$\square + 8 = 17$

Name ______________________

Fact Homework 58B

***Saxon** Math 2 (for use with **Lesson 58**)*

Set 11: Sums of 15, 16, 17, and 18; Review Facts

A. 1. Read the answers to someone.
2. Write the answers.
3. Ask someone to correct your paper. Corrected by ______________________

$6 + 9 =$ ____ $7 + 5 =$ ____ $8 + 6 =$ ____

$8 + 5 =$ ____ $9 + 7 =$ ____ $5 + 3 =$ ____

$5 + 9 =$ ____ $4 + 8 =$ ____ $9 + 8 =$ ____

$5 + 8 =$ ____ $8 + 7 =$ ____ $3 + 7 =$ ____

$7 + 9 =$ ____ $3 + 8 =$ ____ $9 + 4 =$ ____

$7 + 8 =$ ____ $6 + 3 =$ ____ $9 + 6 =$ ____

$6 + 8 =$ ____ $8 + 9 =$ ____ $5 + 7 =$ ____

B. Fill in the missing numbers.

$6 + \square = 15$ $7 + \square = 15$ $8 + \square = 15$

$\square + 7 = 16$ $\square + 8 = 16$ $\square + 9 = 17$

Name ______________________

Date ______________________

Guided Class Practice 58A

*Saxon Math 2 (for use with **Lesson 58**)*

1. Four children chose red apples, six children chose green apples, and five children chose oranges. How many children chose apples?

Number sentence ______________________

Answer ______________________

2. Circle the numbers that are between 5 and 15.
Put an X on the numbers that are not between 5 and 15.

3 19 11 20 7

3. Someone drew a line of symmetry in each shape. Circle the shape with the incorrect line of symmetry.

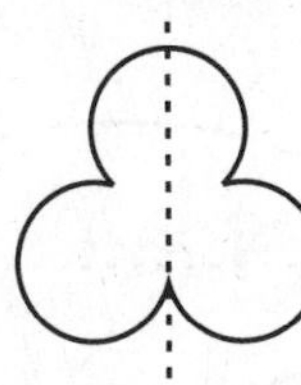 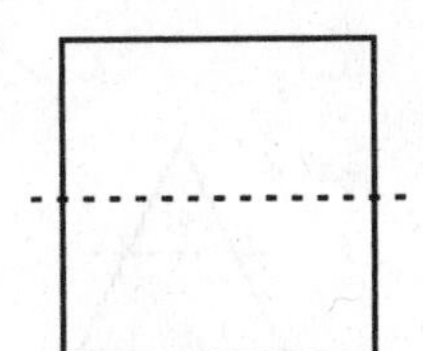 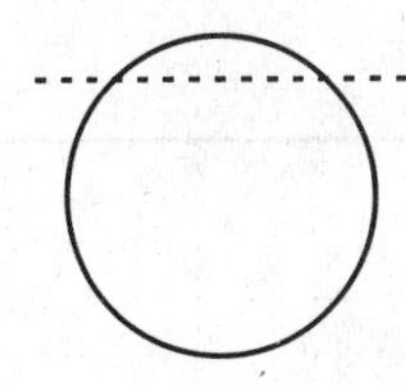 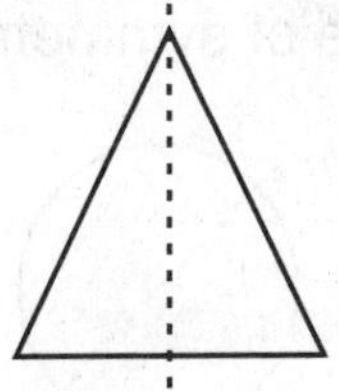

4. Write October 31, 2006 using digits. ______________________

Write the full date for 1/2/04. ______________________

5. Circle the shape that has one half shaded.

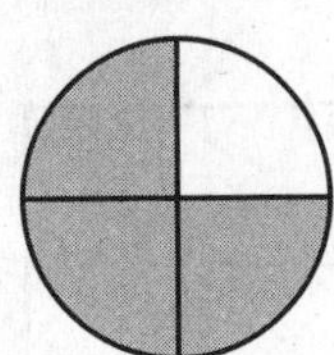 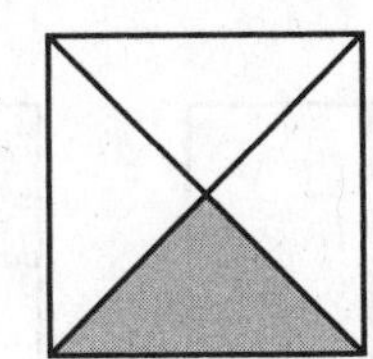 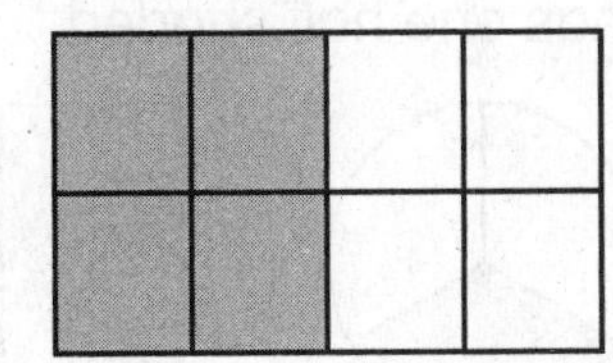 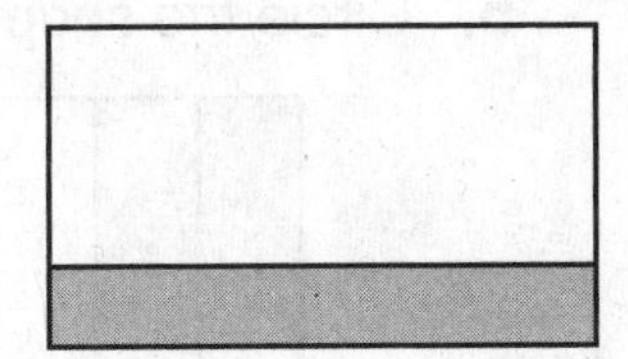

6. Find the sums. Look for 10's.

$8 + 1 + 7 + 9 =$ ______

$6 + 3 + 4 + 5 =$ ______

$$\begin{array}{r} 3 \\ 4 \\ 7 \\ +\ 1 \\ \hline \end{array}$$

Name ______________________________

Date ______________________________

1. Seven children chose chocolate chip cookies, three children chose chocolate ice cream, and eight children chose peanut butter cookies. How many children chose cookies?

Number sentence ______________________________

Answer ______________________________

2. Circle the numbers that are between 3 and 13.
Put an X on the numbers that are not between 3 and 13.

8 2 17 6 12

3. Someone drew a line of symmetry in each shape. Circle the shape with the incorrect line of symmetry.

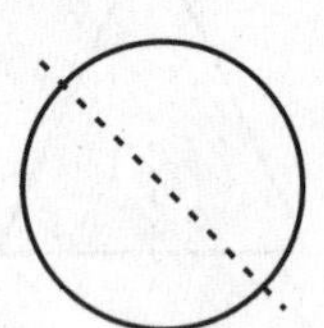
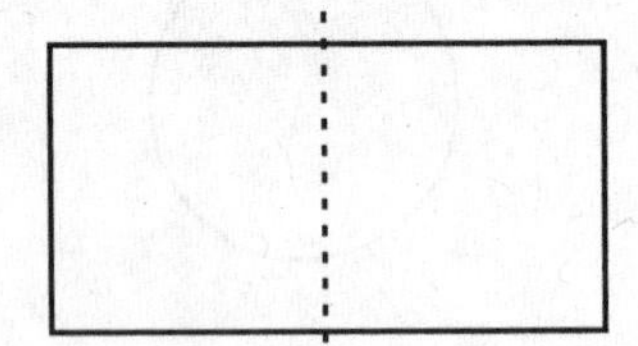
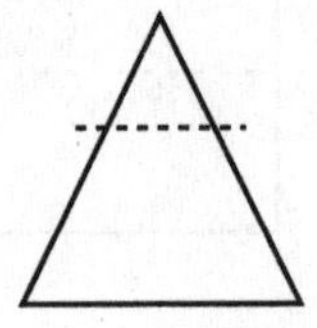
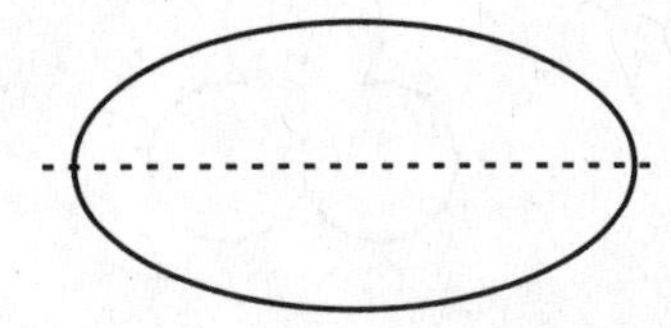

4. Write June 20, 2005 using digits. ______________

Write the full date for 5/9/20. ______________________________

5. Circle the shape that has one half shaded.

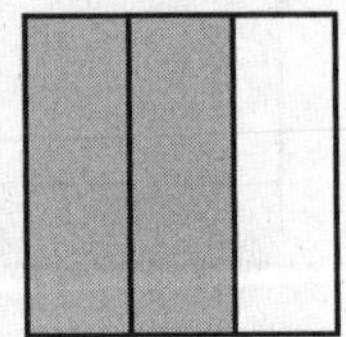
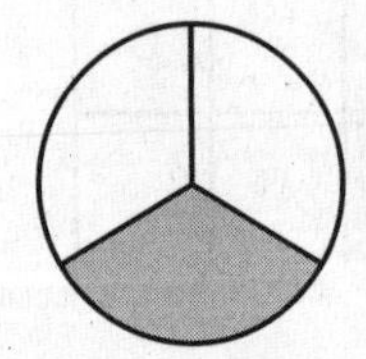
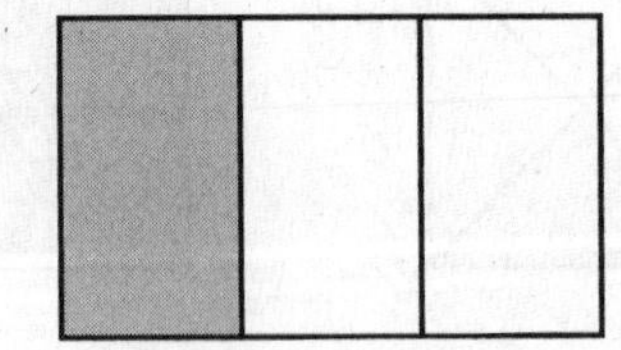
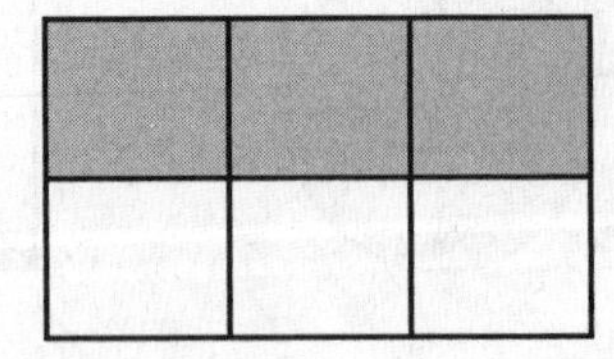

6. Find the sums. Look for 10's.

$7 + 3 + 2 + 1 =$ ______ $5 + 5 + 4 + 3 =$ ______

$$\begin{array}{r} 7 \\ 7 \\ +\ 3 \\ \hline \end{array}$$

Name ______________________

Class Fact Practice 59A

***Saxon** Math 2 (for use with **Lesson 59**)*

Set 11: Sums of 15, 16, 17, and 18; Review Facts

5 + 7	8 + 9	6 + 3	9 + 6	3 + 8
6 + 8	9 + 4	7 + 8	3 + 7	9 + 8
4 + 8	9 + 7	5 + 8	4 + 3	8 + 7
9 + 9	5 + 3	8 + 5	7 + 9	8 + 6
7 + 5	6 + 9	4 + 5	5 + 9	8 + 8

Name ______________________________

Fact Homework 59B

***Saxon** Math 2 (for use with **Lesson 59**)*

Set 11: Sums of 15, 16, 17, and 18; Review Facts

1. Read the answers to someone.
2. Write the answers.
3. Ask someone to correct your paper. Corrected by ______________________

6 + 8	7 + 5	8 + 9	5 + 3	8 + 5
9 + 7	6 + 3	7 + 8	5 + 9	9 + 9
6 + 9	4 + 5	8 + 8	9 + 4	3 + 8
8 + 6	9 + 8	4 + 3	8 + 7	4 + 8
7 + 9	5 + 8	3 + 7	9 + 6	5 + 7

Name ______________________________

Date ______________________________

Guided Class Practice 59A

***Saxon** Math 2 (for use with **Lesson 59**)*

1. Raquel had 3 dimes and 4 pennies. Her sister gave her 2 dimes.

How many dimes does she have now? ________

How many pennies does she have now? ________

How much money is this? ____________

2. Label each piece using a fraction.

Color $\frac{1}{2}$ blue.

Color $\frac{1}{4}$ red.

Color $\frac{1}{3}$ yellow.

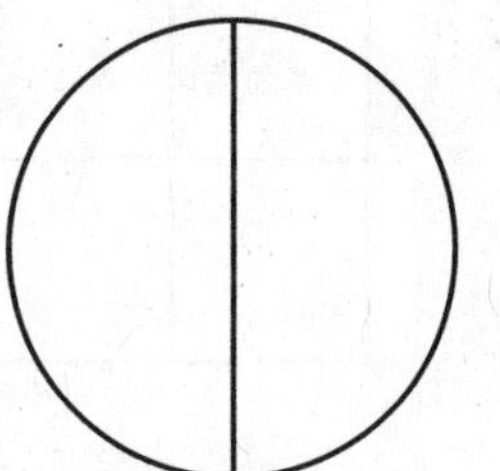
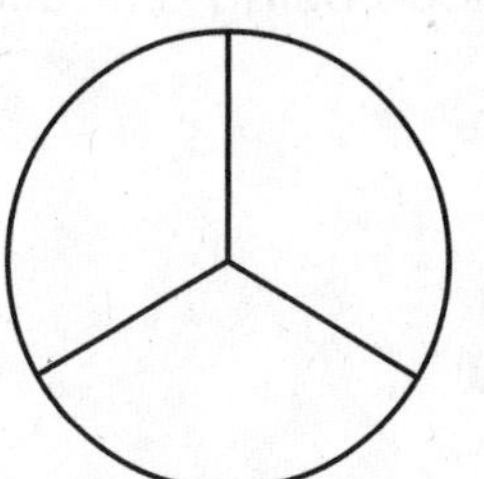
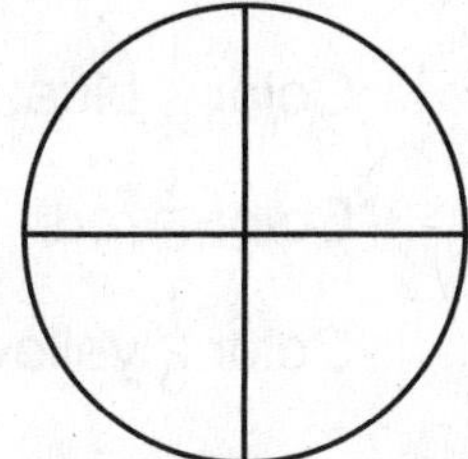

3. Put these numbers in order from least to greatest.

23	47	50	32

________ ________ ________ ________
least greatest

4. Draw 12 shoes.
Circle the pairs.

How many pairs of shoes are there? ________ pairs

5. Name something in your classroom that is about 1 foot long. ________________

Name something in your classroom that is about 3 feet long. ________________

6. Find the sums.

4 + 7 + 1 + 3 = ________

	dimes	pennies	
	2	5	¢
+	3	4	¢
			¢

	dimes	pennies	
	1	7	¢
+	7	2	¢
			¢

Name ____________________

Date ____________________

1. Willie had 6 dimes and 5 pennies. His sister gave him 3 dimes.

How many dimes does he have now? ________

How many pennies does he have now? ________

How much money is this? ____________

2. Label each piece using a fraction.

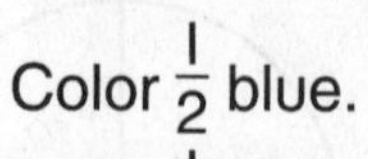

Color $\frac{1}{2}$ blue.

Color $\frac{1}{4}$ red.

Color $\frac{1}{3}$ yellow.

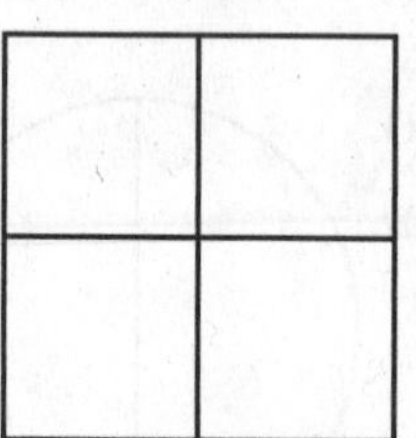
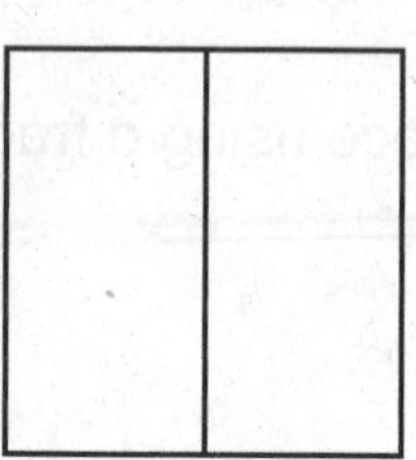
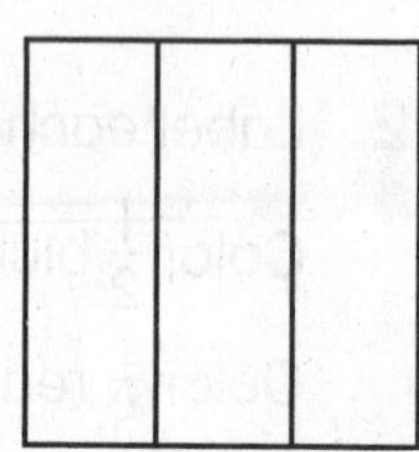

3. Put these numbers in order from least to greatest.

35	41	17	58

________ ________ ________ ________
least greatest

4. Draw 14 earrings.
Circle the pairs.

How many pairs of earrings are there? ________

5. Use your feet to measure the length of your bedroom.

(Walk in a straight line.) ________ feet

Have someone else in your family measure your bedroom with his or her feet.

Name ____________________ ________ feet

Who took more steps? ____________________

Why? ____

6. Find the sums.

$3 + 5 + 9 + 5 + 1 =$ ________

	dimes	pennies	
	3	5	¢
+	4	1	¢
			¢

	dimes	pennies	
	6	3	¢
+	2	3	¢
			¢

Name ______________________ Score ______

Fact Assessment 11

***Saxon** Math 2 (for use with **Lesson 60-1**)*

Set 11: Sums of 15, 16, 17, and 18; Review Facts

$6 + 9$	$3 + 7$	$5 + 8$	$8 + 9$	$6 + 8$
$9 + 4$	$7 + 5$	$8 + 7$	$6 + 3$	$7 + 9$
$8 + 5$	$9 + 8$	$4 + 5$	$7 + 8$	$5 + 3$
$9 + 6$	$3 + 8$	$5 + 7$	$9 + 9$	$8 + 6$
$4 + 3$	$9 + 7$	$4 + 8$	$5 + 9$	$8 + 8$

Name ______________________________

Date ______________________________

Problem-Solving Worksheet 60A

***Saxon** Math 2 (for use with **Lesson 60-1**)*

Understand	Plan	Solve	Check

Draw a Picture

Look for a Pattern

Courtney wants to make a bracelet with a yellow, yellow, red, green, green repeating pattern. She will use 10 beads to make her bracelet. Show how many of each color bead she will need.

How many yellow, red, and green beads will Courtney need to make her bracelet? _______ yellow _______ red _______ green

Name ____________________

Performance Task Worksheet 60B

***Saxon** Math 2 (for use with **Lesson 60-1**)*

Understand	Plan	Solve	Check

DeAnna wants to make a necklace with a blue, yellow, red, yellow repeating pattern. She will use 12 beads to make her necklace. Show how many of each color bead she will need.

How many blue, yellow, and red beads will DeAnna need to make her necklace? _______ blue _______ yellow _______ red

Circle the problem-solving strategies you used to solve this problem.

Act It Out ***Use Logical Reasoning***

Draw a Picture ***Look for a Pattern***

Make an Organized List

Explain how you got your answer: ____________________

____

____

____

Name ______________________

Date ______________________

1. The children in Mrs. Rafone's class had a picnic. Each child chose a hot dog or a hamburger for lunch. Sixteen children ate hamburgers and ten children ate hot dogs. How many children were at the picnic?

Number sentence ______________________

Answer ______________________

2. How many tally marks are shown? ________

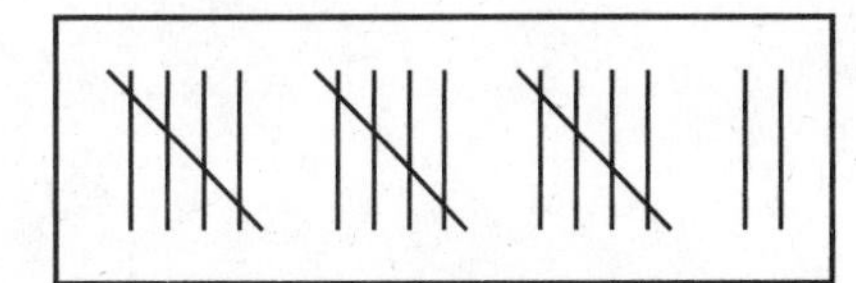

Show **28** using tally marks.

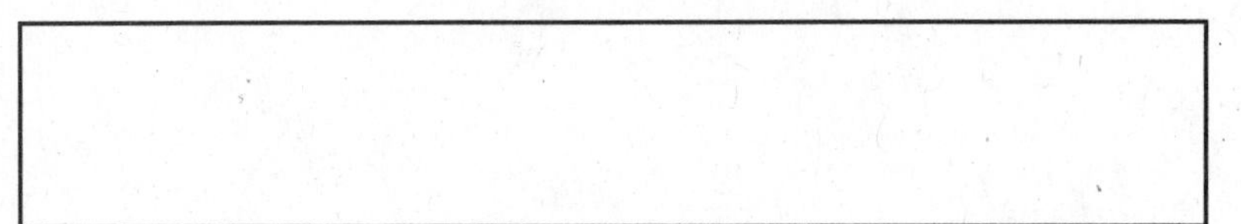

3. How much money is this? ________

4. Color three fourths green.
Color five eighths yellow.
Color one half blue.

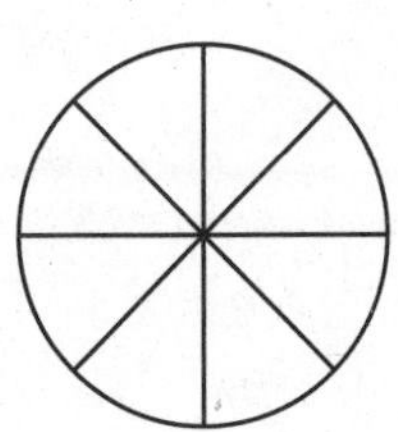
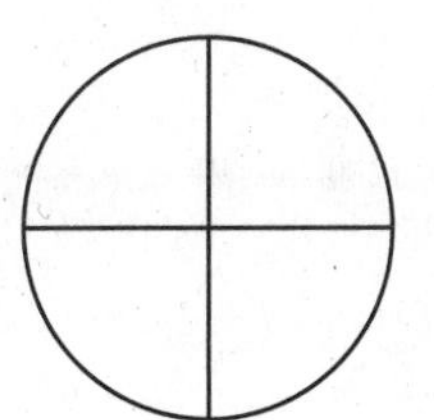
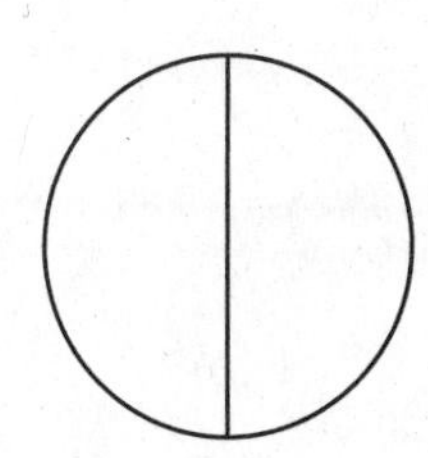

5. Circle the largest number. Put an X on the smallest number. Write the numbers in order from least to greatest.

47 45 52

________ ________ ________
least greatest

6. Write August 23, 2015 using digits. ________

Write the full date for 2/3/05. ______________________

Name ______________________________

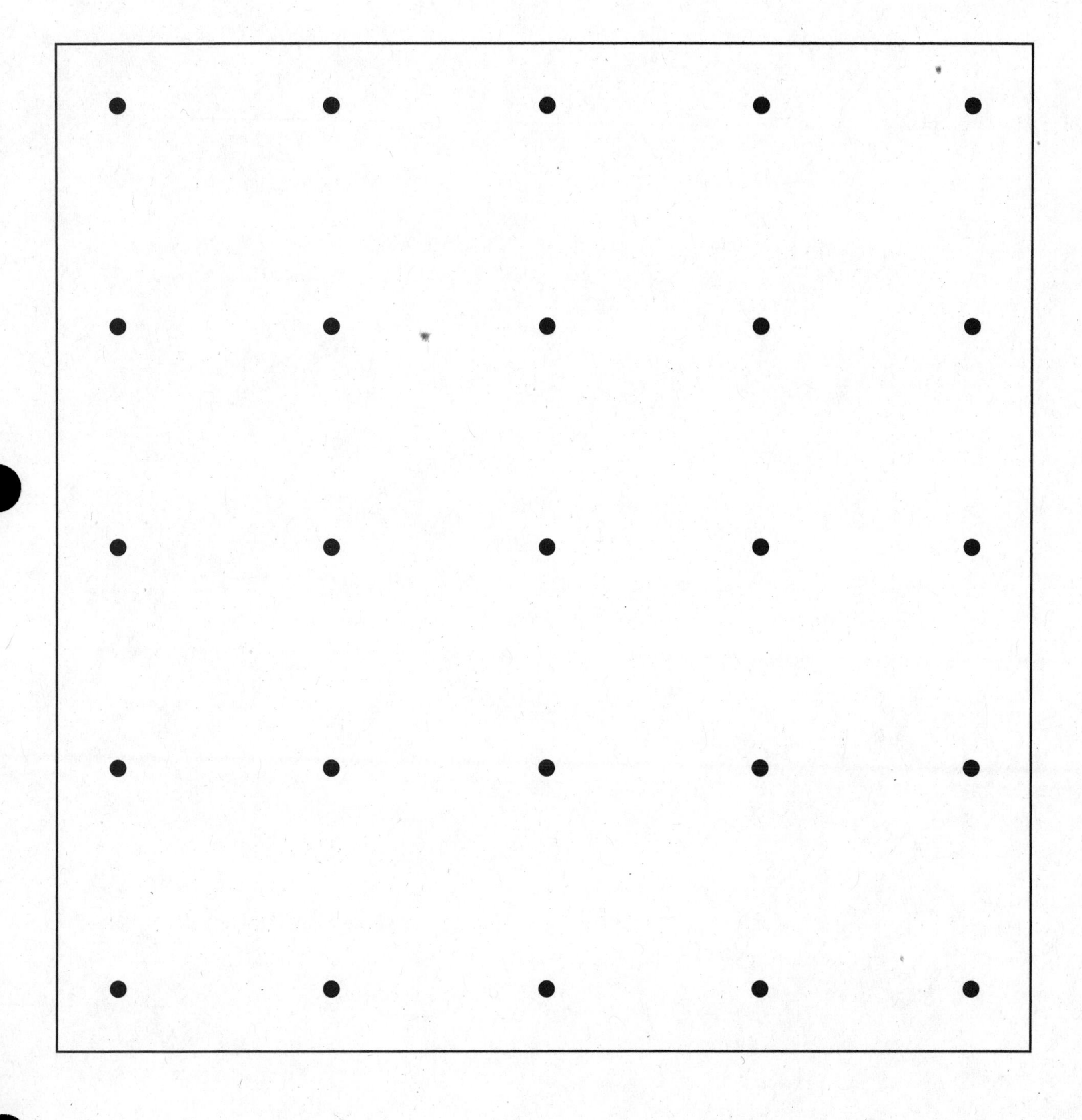

Name ______________________________

Class Fact Practice 61A

***Saxon** Math 2 (for use with **Lesson 61**)*

Set 12: 100 Addition Facts

1	9 + 1	2 + 2	6 + 4	5 + 1	0 + 7	9 + 9	7 + 3	1 + 6	2 + 5	5 + 4
2	9 + 4	2 + 0	8 + 7	4 + 1	6 + 6	7 + 8	3 + 2	9 + 8	0 + 8	4 + 6
3	5 + 2	3 + 9	0 + 6	8 + 1	3 + 3	7 + 4	7 + 0	1 + 5	6 + 7	2 + 3
4	1 + 0	5 + 5	7 + 6	3 + 4	2 + 1	9 + 5	7 + 2	4 + 9	0 + 3	6 + 8
5	8 + 2	3 + 5	1 + 7	0 + 0	6 + 2	5 + 7	1 + 4	8 + 6	2 + 9	5 + 0
6	6 + 3	0 + 5	3 + 7	4 + 4	9 + 2	1 + 8	6 + 5	2 + 4	8 + 8	0 + 9
7	4 + 2	7 + 7	9 + 0	9 + 6	5 + 8	0 + 1	3 + 6	7 + 9	6 + 0	4 + 8
8	7 + 1	2 + 6	4 + 7	1 + 2	4 + 5	8 + 9	3 + 0	8 + 3	1 + 9	5 + 6
9	1 + 1	3 + 8	0 + 2	5 + 9	9 + 3	2 + 7	8 + 0	4 + 3	6 + 9	1 + 3
10	8 + 5	4 + 0	5 + 3	2 + 8	3 + 1	7 + 5	9 + 7	0 + 4	8 + 4	6 + 1

Name ________________________________

Fact Homework 61B

Saxon Math 2 (for use with Lesson 61)

Set 12: 100 Addition Facts   Corrected by ______

1	9 + 1	2 + 2	6 + 4	5 + 1	0 + 7	9 + 9	7 + 3	1 + 6	2 + 5	5 + 4
2	9 + 4	2 + 0	8 + 7	4 + 1	6 + 6	7 + 8	3 + 2	9 + 8	0 + 8	4 + 6
3	5 + 2	3 + 9	0 + 6	8 + 1	3 + 3	7 + 4	7 + 0	1 + 5	6 + 7	2 + 3
4	1 + 0	5 + 5	7 + 6	3 + 4	2 + 1	9 + 5	7 + 2	4 + 9	0 + 3	6 + 8
5	8 + 2	3 + 5	1 + 7	0 + 0	6 + 2	5 + 7	1 + 4	8 + 6	2 + 9	5 + 0
6	6 + 3	0 + 5	3 + 7	4 + 4	9 + 2	1 + 8	6 + 5	2 + 4	8 + 8	0 + 9
7	4 + 2	7 + 7	9 + 0	9 + 6	5 + 8	0 + 1	3 + 6	7 + 9	6 + 0	4 + 8
8	7 + 1	2 + 6	4 + 7	1 + 2	4 + 5	8 + 9	3 + 0	8 + 3	1 + 9	5 + 6
9	1 + 1	3 + 8	0 + 2	5 + 9	9 + 3	2 + 7	8 + 0	4 + 3	6 + 9	1 + 3
10	8 + 5	4 + 0	5 + 3	2 + 8	3 + 1	7 + 5	9 + 7	0 + 4	8 + 4	6 + 1

Name ______________________

Date ______________________

1. Gina has 4 dimes. How much money does she have? ________

Allison has 3 dimes. How much money does she have? ________

How much money do the two girls have altogether?

Number sentence ______________________

Answer ______________________

2. Draw a 5-inch line segment.

•

3. Find the sums.

1 + 7 + 9 + 2 = ________

5 + 6 + 2 + 4 + 3 = ________

	dimes	pennies	
	4	7	¢
+	5	1	¢
			¢

	dimes	pennies	
	1	3	¢
+	7	2	¢
			¢

4. Pretend you are the teacher. Circle the mistakes with a crayon and write the correct answers.

1. 37 + 10 = 47

2. Divide the circle into fourths. Shade one fourth.

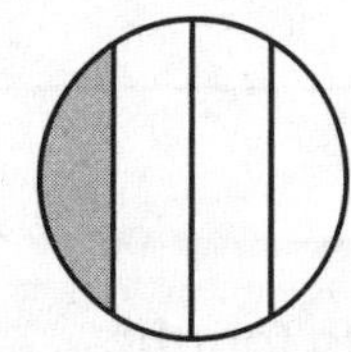

3. Show **23** using tally marks.

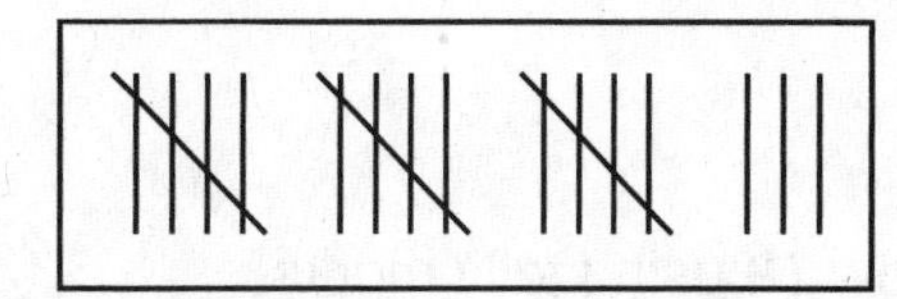

4. Put the numbers in order from least to greatest.

24	37	32
least		greatest

Name ___________________________

Date ___________________________

Homework 61B

***Saxon** Math 2 (for use with **Lesson 61**)*

1. Calvin has 2 dimes. How much money does he have? __________

Tim has 1 dime. How much money does he have? __________

How much money do the two boys have altogether?

Number sentence ___________________________

Answer ___________________________

2. How much money is this? __________

3. Find the sums.

2 + 5 + 3 + 8 = ______

4 + 1 + 7 + 3 + 6 = ______

	dimes	pennies	
	4	5	¢
+	1	3	¢
			¢

	dimes	pennies	
	3	6	¢
+	5	2	¢
			¢

4. Pretend you are the teacher. Circle the mistakes with a crayon and write the correct answers.

1. 25 + 10 = 15

2. Divide the square into fourths. Shade three fourths.

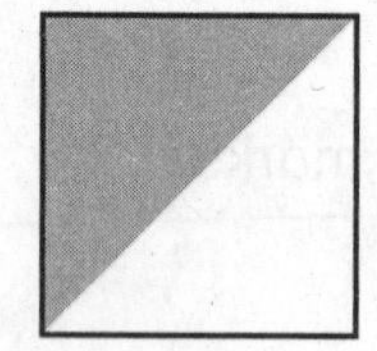

3. Show **14** using tally marks.

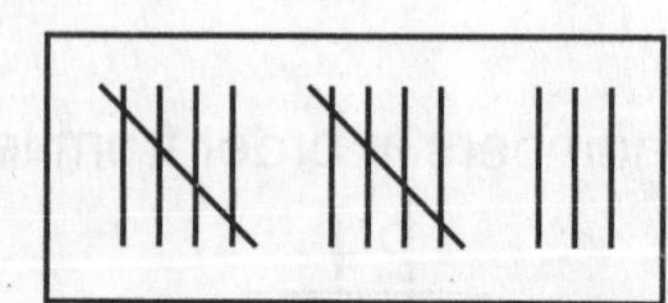

4. Put the numbers in order from least to greatest.

58	43	31
least		greatest

Name ______________________

Class Fact Practice 62A

***Saxon** Math 2 (for use with **Lesson 62**)*

Set 12: 100 Addition Facts

1	9 + 1	2 + 2	6 + 4	5 + 1	0 + 7	9 + 9	7 + 3	1 + 6	2 + 5	5 + 4
2	9 + 4	2 + 0	8 + 7	4 + 1	6 + 6	7 + 8	3 + 2	9 + 8	0 + 8	4 + 6
3	5 + 2	3 + 9	0 + 6	8 + 1	3 + 3	7 + 4	7 + 0	1 + 5	6 + 7	2 + 3
4	1 + 0	5 + 5	7 + 6	3 + 4	2 + 1	9 + 5	7 + 2	4 + 9	0 + 3	6 + 8
5	8 + 2	3 + 5	1 + 7	0 + 0	6 + 2	5 + 7	1 + 4	8 + 6	2 + 9	5 + 0
6	6 + 3	0 + 5	3 + 7	4 + 4	9 + 2	1 + 8	6 + 5	2 + 4	8 + 8	0 + 9
7	4 + 2	7 + 7	9 + 0	9 + 6	5 + 8	0 + 1	3 + 6	7 + 9	6 + 0	4 + 8
8	7 + 1	2 + 6	4 + 7	1 + 2	4 + 5	8 + 9	3 + 0	8 + 3	1 + 9	5 + 6
9	1 + 1	3 + 8	0 + 2	5 + 9	9 + 3	2 + 7	8 + 0	4 + 3	6 + 9	1 + 3
10	8 + 5	4 + 0	5 + 3	2 + 8	3 + 1	7 + 5	9 + 7	0 + 4	8 + 4	6 + 1

Name ______________________________

Fact Homework 62B

***Saxon** Math 2 (for use with **Lesson 62**)*

Set 12: 100 Addition Facts   Corrected by ________

1	9 + 1	2 + 2	6 + 4	5 + 1	0 + 7	9 + 9	7 + 3	1 + 6	2 + 5	5 + 4
2	9 + 4	2 + 0	8 + 7	4 + 1	6 + 6	7 + 8	3 + 2	9 + 8	0 + 8	4 + 6
3	5 + 2	3 + 9	0 + 6	8 + 1	3 + 3	7 + 4	7 + 0	1 + 5	6 + 7	2 + 3
4	1 + 0	5 + 5	7 + 6	3 + 4	2 + 1	9 + 5	7 + 2	4 + 9	0 + 3	6 + 8
5	8 + 2	3 + 5	1 + 7	0 + 0	6 + 2	5 + 7	1 + 4	8 + 6	2 + 9	5 + 0
6	6 + 3	0 + 5	3 + 7	4 + 4	9 + 2	1 + 8	6 + 5	2 + 4	8 + 8	0 + 9
7	4 + 2	7 + 7	9 + 0	9 + 6	5 + 8	0 + 1	3 + 6	7 + 9	6 + 0	4 + 8
8	7 + 1	2 + 6	4 + 7	1 + 2	4 + 5	8 + 9	3 + 0	8 + 3	1 + 9	5 + 6
9	1 + 1	3 + 8	0 + 2	5 + 9	9 + 3	2 + 7	8 + 0	4 + 3	6 + 9	1 + 3
10	8 + 5	4 + 0	5 + 3	2 + 8	3 + 1	7 + 5	9 + 7	0 + 4	8 + 4	6 + 1

Name ____________________

Dimes	Pennies

1.

	dimes	pennies	
	4	3	¢
+	2	5	¢
			¢

2.

	dimes	pennies	
	6	6	¢
+	1	8	¢
			¢

3.

	dimes	pennies	
	4	9	¢
+	3	7	¢
			¢

4.

  4 4 ¢
+ 4 6 ¢
_____
  ¢

5.

  6 5 ¢
+ 2 2 ¢
_____
  ¢

6.

  8 ¢
+ 5 3 ¢
_____
  ¢

Name ______________________

Date ______________________

Guided Class Practice 62A

***Saxon** Math 2 (for use with **Lesson 62**)*

1. Seventeen children were on the bus. Ten children got on at the third stop. How many children are on the bus now?

Number sentence ______________________

Answer ______________________

2. Finish numbering the number line using the even numbers.

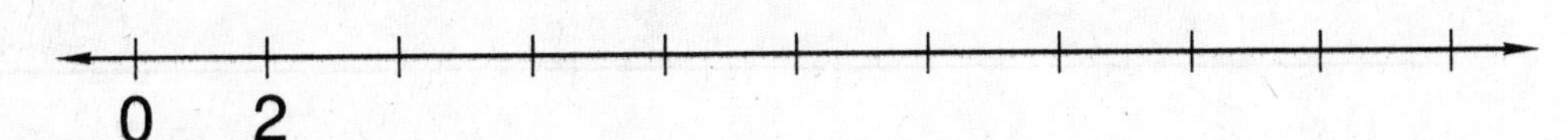

Put a point at **12.** Label it **A.**
Put a point at **8.** Label it **B.**

3. Match each fraction with the correct picture.

$\frac{1}{4}$ •

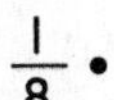

$\frac{1}{8}$ •

$\frac{1}{3}$ •

$\frac{1}{2}$ •

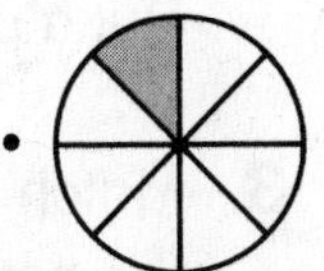

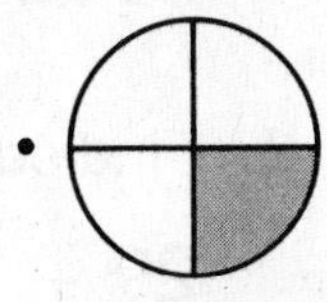

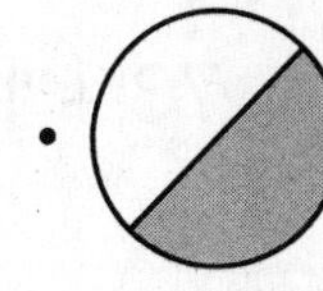

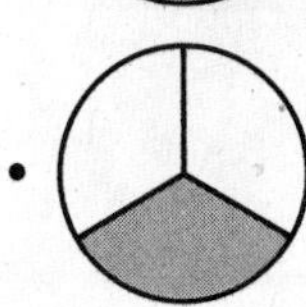

4. Find each answer.

$46 + 10 =$ ______

$10 + 27 =$ ______

5. Circle the shape that is congruent to the shape on the left.
(Congruent means same shape and same size.)

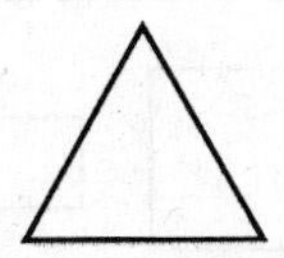

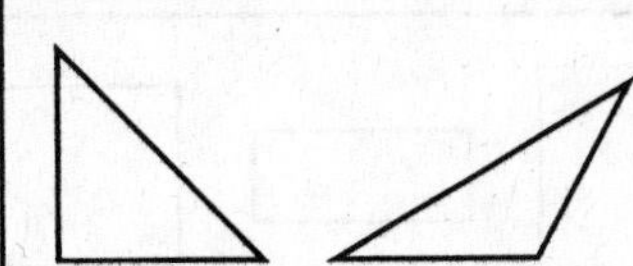

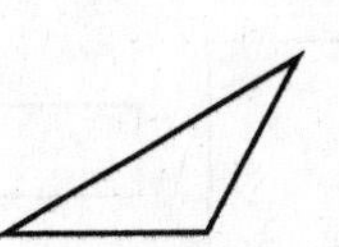

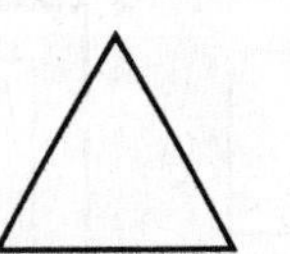

6. Use dimes and pennies to find the sums.

	dimes	pennies	
	5	4	¢
+	2	8	¢
			¢

	dimes	pennies	
	3	9	¢
+	2	1	¢
			¢

	dimes	pennies	
	2	4	¢
+	6	3	¢
			¢

Name ______________________

Date ______________________

Homework 62B

***Saxon** Math 2 (for use with **Lesson 62**)*

1. Twenty-seven children were on the bus. Ten children got on at the fourth stop. How many children are on the bus now?

Number sentence ______________________

Answer ______________________

2. Finish numbering the number line using the even numbers.

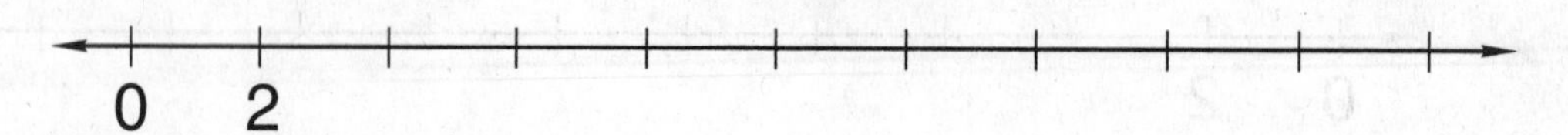

Put a point at **14.** Label it **C.**
Put a point at **6.** Label it **D.**

3. Match each fraction with the correct picture.

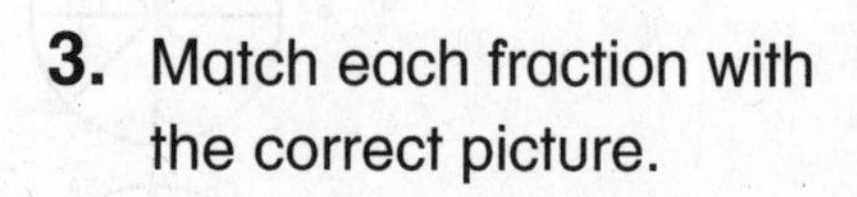

$\frac{1}{4}$ •

$\frac{1}{3}$ •

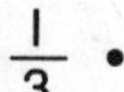

$\frac{1}{8}$ •

$\frac{1}{2}$ •

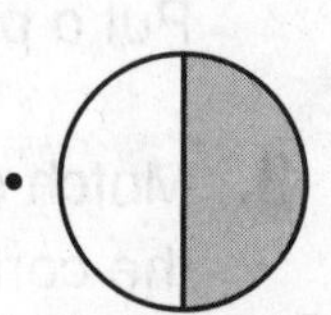

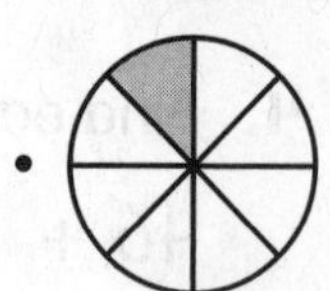

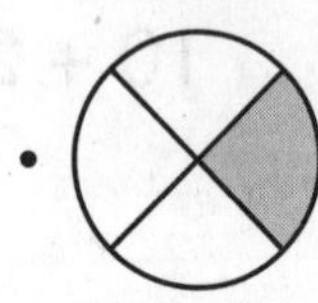

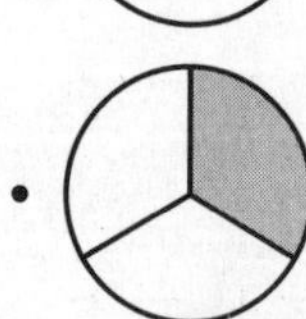

4. Find each answer.

29 + 10 = ______

73 + 10 = ______

5. Circle the shape that is congruent to the shape on the left. (Congruent means same shape and same size.)

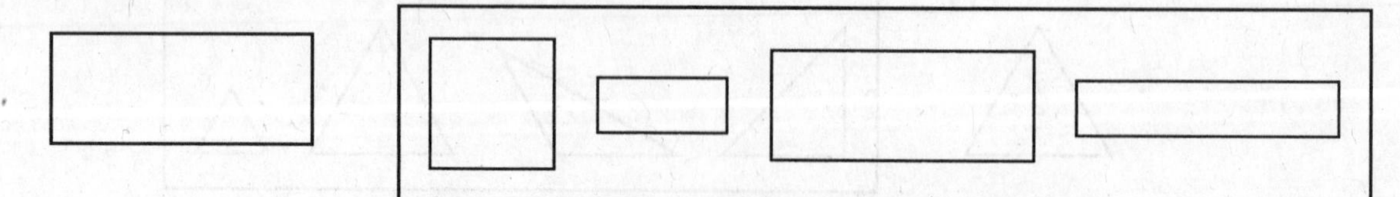

6. Use dimes and pennies to find the sums.

	dimes	pennies	
	5	4	¢
+	1	5	¢
			¢

	dimes	pennies	
	2	8	¢
+	1	3	¢
			¢

	dimes	pennies	
	2	6	¢
+	5	4	¢
			¢

Name ______________________________

Class Fact Practice 63A

***Saxon** Math 2 (for use with **Lesson 63**)*

Set 12: 100 Addition Facts

1	9 + 1	2 + 2	6 + 4	5 + 1	0 + 7	9 + 9	7 + 3	1 + 6	2 + 5	5 + 4
2	9 + 4	2 + 0	8 + 7	4 + 1	6 + 6	7 + 8	3 + 2	9 + 8	0 + 8	4 + 6
3	5 + 2	3 + 9	0 + 6	8 + 1	3 + 3	7 + 4	7 + 0	1 + 5	6 + 7	2 + 3
4	1 + 0	5 + 5	7 + 6	3 + 4	2 + 1	9 + 5	7 + 2	4 + 9	0 + 3	6 + 8
5	8 + 2	3 + 5	1 + 7	0 + 0	6 + 2	5 + 7	1 + 4	8 + 6	2 + 9	5 + 0
6	6 + 3	0 + 5	3 + 7	4 + 4	9 + 2	1 + 8	6 + 5	2 + 4	8 + 8	0 + 9
7	4 + 2	7 + 7	9 + 0	9 + 6	5 + 8	0 + 1	3 + 6	7 + 9	6 + 0	4 + 8
8	7 + 1	2 + 6	4 + 7	1 + 2	4 + 5	8 + 9	3 + 0	8 + 3	1 + 9	5 + 6
9	1 + 1	3 + 8	0 + 2	5 + 9	9 + 3	2 + 7	8 + 0	4 + 3	6 + 9	1 + 3
10	8 + 5	4 + 0	5 + 3	2 + 8	3 + 1	7 + 5	9 + 7	0 + 4	8 + 4	6 + 1

Name ______________________

Fact Homework 63B

***Saxon** Math 2 (for use with **Lesson 63**)*

Set 12: 100 Addition Facts   Corrected by ______

1	9 + 1	2 + 2	6 + 4	5 + 1	0 + 7	9 + 9	7 + 3	1 + 6	2 + 5	5 + 4
2	9 + 4	2 + 0	8 + 7	4 + 1	6 + 6	7 + 8	3 + 2	9 + 8	0 + 8	4 + 6
3	5 + 2	3 + 9	0 + 6	8 + 1	3 + 3	7 + 4	7 + 0	1 + 5	6 + 7	2 + 3
4	1 + 0	5 + 5	7 + 6	3 + 4	2 + 1	9 + 5	7 + 2	4 + 9	0 + 3	6 + 8
5	8 + 2	3 + 5	1 + 7	0 + 0	6 + 2	5 + 7	1 + 4	8 + 6	2 + 9	5 + 0
6	6 + 3	0 + 5	3 + 7	4 + 4	9 + 2	1 + 8	6 + 5	2 + 4	8 + 8	0 + 9
7	4 + 2	7 + 7	9 + 0	9 + 6	5 + 8	0 + 1	3 + 6	7 + 9	6 + 0	4 + 8
8	7 + 1	2 + 6	4 + 7	1 + 2	4 + 5	8 + 9	3 + 0	8 + 3	1 + 9	5 + 6
9	1 + 1	3 + 8	0 + 2	5 + 9	9 + 3	2 + 7	8 + 0	4 + 3	6 + 9	1 + 3
10	8 + 5	4 + 0	5 + 3	2 + 8	3 + 1	7 + 5	9 + 7	0 + 4	8 + 4	6 + 1

Name ______________________________

Date ______________________________

1. One of these is my favorite day of the week. Cross out the days of the week that cannot be my favorite day of the week.

It is not the sixth day of the week.
It is not a weekend day.
It does not begin with a T.
It is not the day in the middle of the week.

What is my favorite day? ______________

Sunday
Monday
Tuesday
Wednesday
Thursday
Friday
Saturday

2. Put these numbers in order from least to greatest.

49 47 43 36

_______ least _______ _______ _______ greatest

3. Put a dot inside each angle. Count the number of angles in each shape.

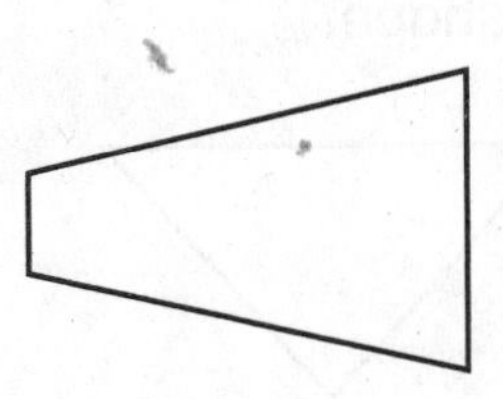

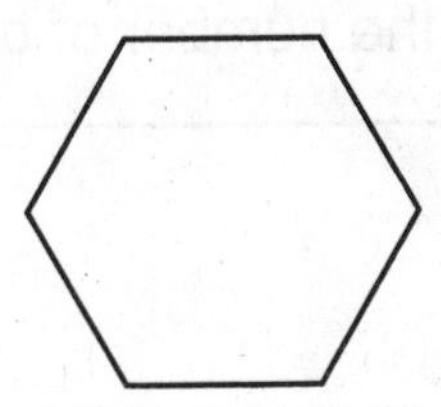

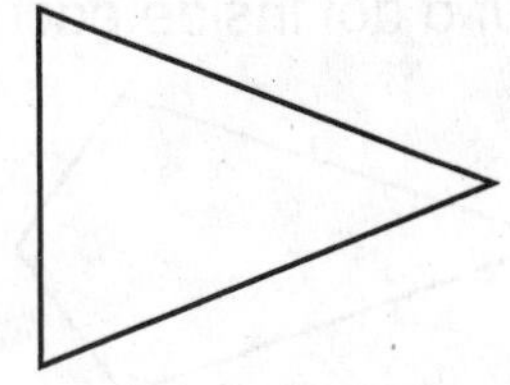

_______ angles _______ angles _______ angles

4. Draw a line of symmetry in each shape in Problem 3.

5. Write the fraction that tells what part of each set is shaded.

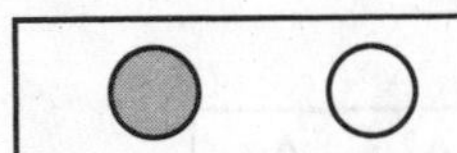

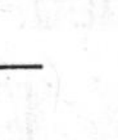

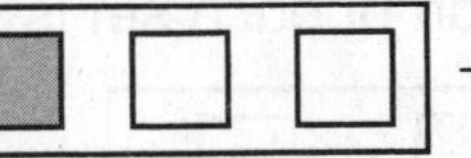

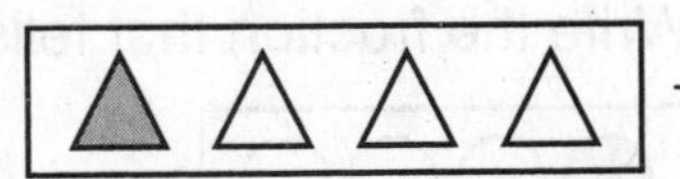

6. Draw a 4-inch line segment.

•

7. Find the sums.

$2 + 7 + 1 + 4 + 3 =$ _______

$6 + 5 + 1 + 5 + 4 =$ _______

Name ______________________

Date ______________________

Homework 63B

***Saxon** Math 2 (for use with **Lesson 63**)*

1. One of these is my brother's favorite day of the week. Cross out the days of the week that cannot be my brother's favorite day of the week.

It is not the first or third day of the week.
It does not have exactly 6 letters.
It is not the last day of the week.
It is not the day in the middle of the week.

What is my brother's favorite day? ______________

Sunday
Monday
Tuesday
Wednesday
Thursday
Friday
Saturday

2. Put these numbers in order from least to greatest.

56 54 59 51

______ ______ ______ ______
least ... greatest

3. Put a dot inside each angle. Count the number of angles in each shape.

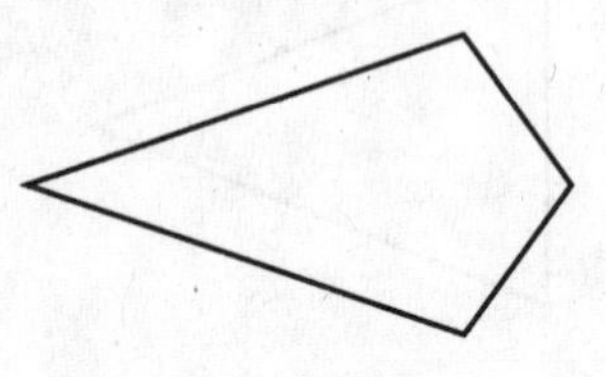

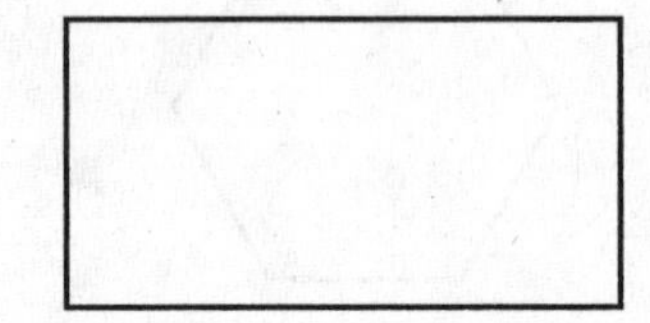

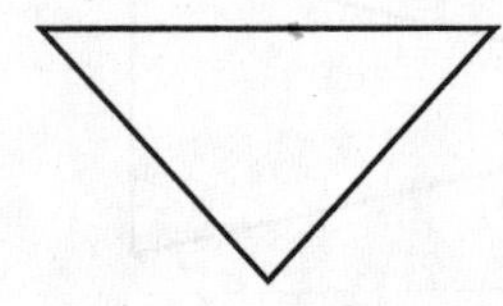

______ angles ______ angles ______ angles

4. Draw a line of symmetry in each shape in Problem 3.

5. Write the fraction that tells what part of each set is shaded.

 ______

 ______ 

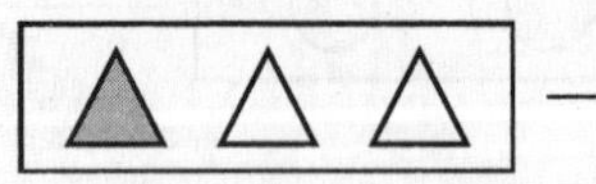 ______

6. Write December 5, 2007 using digits. ______________

7. Find the sums.

$4 + 3 + 6 + 2 + 7 =$ ______

$4 + 1 + 4 + 6 + 6 =$ ______

Name ______________________________

Class Fact Practice 64A

***Saxon** Math 2 (for use with **Lesson 64**)*

Set 12: 100 Addition Facts

1	$9 + 1$	$2 + 2$	$6 + 4$	$5 + 1$	$0 + 7$	$9 + 9$	$7 + 3$	$1 + 6$	$2 + 5$	$5 + 4$
2	$9 + 4$	$2 + 0$	$8 + 7$	$4 + 1$	$6 + 6$	$7 + 8$	$3 + 2$	$9 + 8$	$0 + 8$	$4 + 6$
3	$5 + 2$	$3 + 9$	$0 + 6$	$8 + 1$	$3 + 3$	$7 + 4$	$7 + 0$	$1 + 5$	$6 + 7$	$2 + 3$
4	$1 + 0$	$5 + 5$	$7 + 6$	$3 + 4$	$2 + 1$	$9 + 5$	$7 + 2$	$4 + 9$	$0 + 3$	$6 + 8$
5	$8 + 2$	$3 + 5$	$1 + 7$	$0 + 0$	$6 + 2$	$5 + 7$	$1 + 4$	$8 + 6$	$2 + 9$	$5 + 0$
6	$6 + 3$	$0 + 5$	$3 + 7$	$4 + 4$	$9 + 2$	$1 + 8$	$6 + 5$	$2 + 4$	$8 + 8$	$0 + 9$
7	$4 + 2$	$7 + 7$	$9 + 0$	$9 + 6$	$5 + 8$	$0 + 1$	$3 + 6$	$7 + 9$	$6 + 0$	$4 + 8$
8	$7 + 1$	$2 + 6$	$4 + 7$	$1 + 2$	$4 + 5$	$8 + 9$	$3 + 0$	$8 + 3$	$1 + 9$	$5 + 6$
9	$1 + 1$	$3 + 8$	$0 + 2$	$5 + 9$	$9 + 3$	$2 + 7$	$8 + 0$	$4 + 3$	$6 + 9$	$1 + 3$
10	$8 + 5$	$4 + 0$	$5 + 3$	$2 + 8$	$3 + 1$	$7 + 5$	$9 + 7$	$0 + 4$	$8 + 4$	$6 + 1$

Name ______________________________

Fact Homework 64B

***Saxon** Math 2 (for use with **Lesson 64**)*

Set 12: 100 Addition Facts   Corrected by ______

1	9 + 1	2 + 2	6 + 4	5 + 1	0 + 7	9 + 9	7 + 3	1 + 6	2 + 5	5 + 4
2	9 + 4	2 + 0	8 + 7	4 + 1	6 + 6	7 + 8	3 + 2	9 + 8	0 + 8	4 + 6
3	5 + 2	3 + 9	0 + 6	8 + 1	3 + 3	7 + 4	7 + 0	1 + 5	6 + 7	2 + 3
4	1 + 0	5 + 5	7 + 6	3 + 4	2 + 1	9 + 5	7 + 2	4 + 9	0 + 3	6 + 8
5	8 + 2	3 + 5	1 + 7	0 + 0	6 + 2	5 + 7	1 + 4	8 + 6	2 + 9	5 + 0
6	6 + 3	0 + 5	3 + 7	4 + 4	9 + 2	1 + 8	6 + 5	2 + 4	8 + 8	0 + 9
7	4 + 2	7 + 7	9 + 0	9 + 6	5 + 8	0 + 1	3 + 6	7 + 9	6 + 0	4 + 8
8	7 + 1	2 + 6	4 + 7	1 + 2	4 + 5	8 + 9	3 + 0	8 + 3	1 + 9	5 + 6
9	1 + 1	3 + 8	0 + 2	5 + 9	9 + 3	2 + 7	8 + 0	4 + 3	6 + 9	1 + 3
10	8 + 5	4 + 0	5 + 3	2 + 8	3 + 1	7 + 5	9 + 7	0 + 4	8 + 4	6 + 1

Name ______________________________

1. 73¢ + 15¢

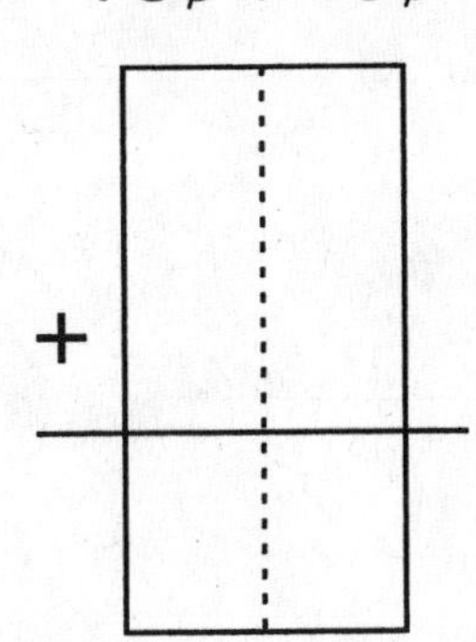

2. 29¢ + 21¢

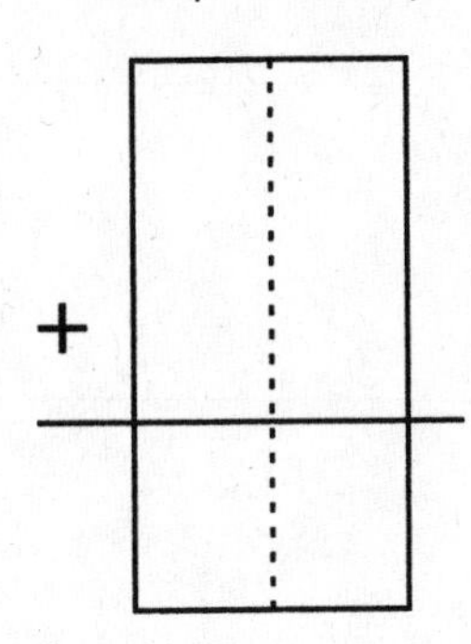

3. 17¢ + 27¢

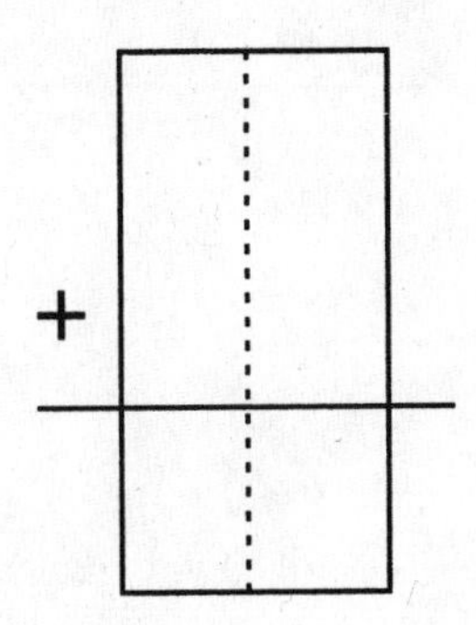

4. 52¢ + 24¢

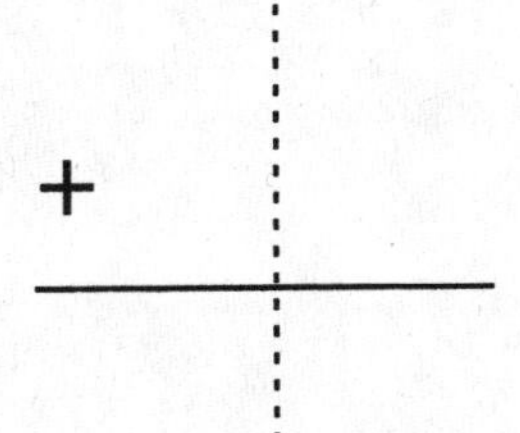

5. 65¢ + 25¢

+

6. 49¢ + 24¢

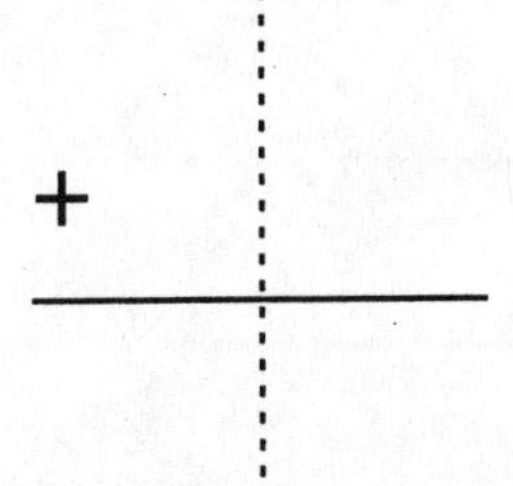

7. 16¢ + 24¢

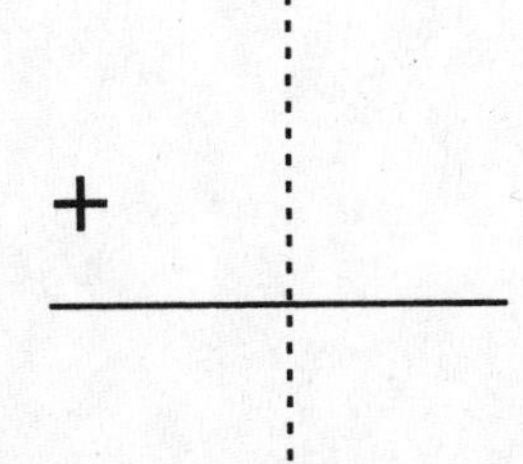

8. 5¢ + 38¢

+

9. 46¢ + 7¢

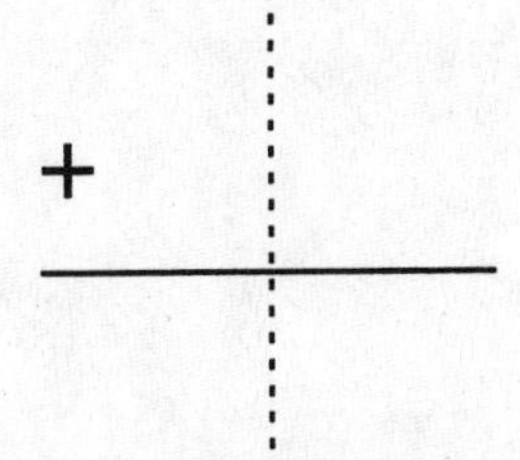

Name ____________________________________

Date ____________________________________

Guided Class Practice 64A

Saxon Math 2 (for use with Lesson 64)

1. Mrs. Reilly wore a different pair of earrings to school each day last week. Draw a picture to show the earrings she wore.

   How many pairs of earrings did she wear?

   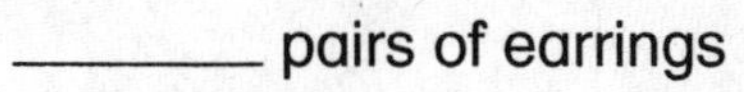
   ________ pairs of earrings

   How many earrings is that? ________ earrings

2. Draw a line of symmetry in each shape.
   Color the shape with 4 angles green.
   Color the shape with 5 angles yellow.
   Color the shape with 3 angles orange.

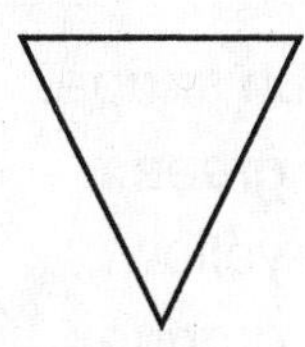
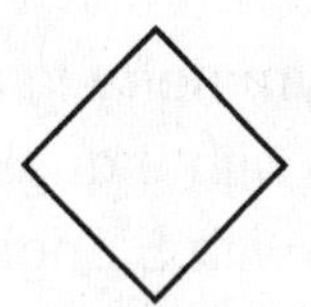
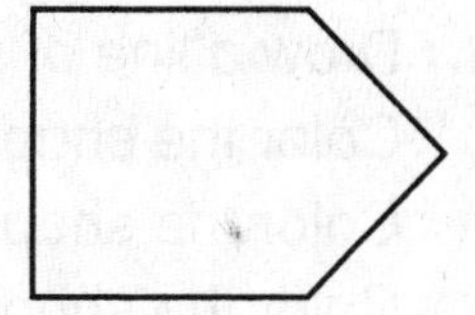

3. Draw 7 pairs of shoes. How many shoes is that? ________

4. Two of these triangles are congruent (same shape and same size). Color the congruent triangles red.

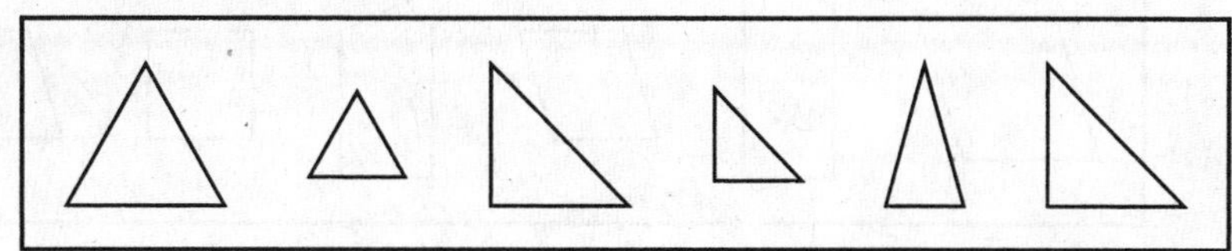

5. About how far is it from the floor to the doorknob on your classroom door?

   12 inches   6 feet   3 feet   2 inches

6. Find the sums.

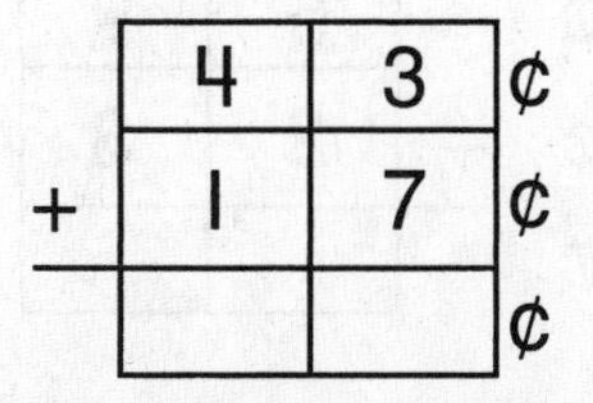

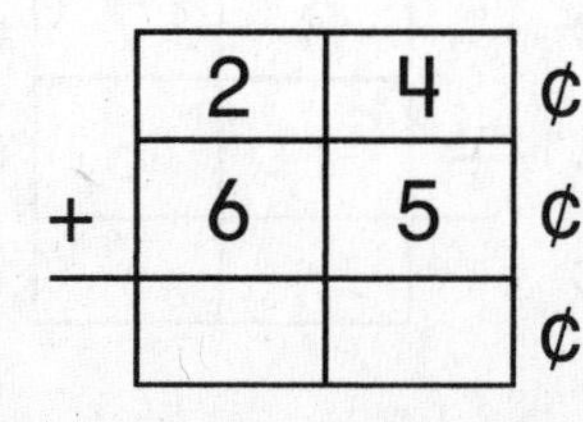

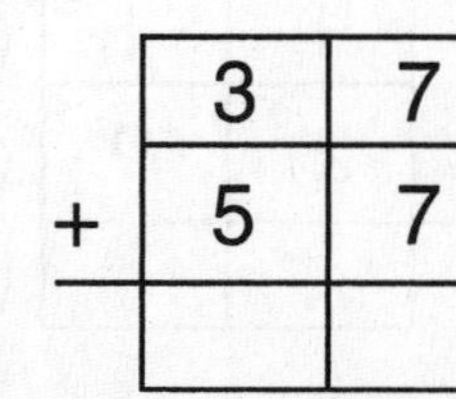

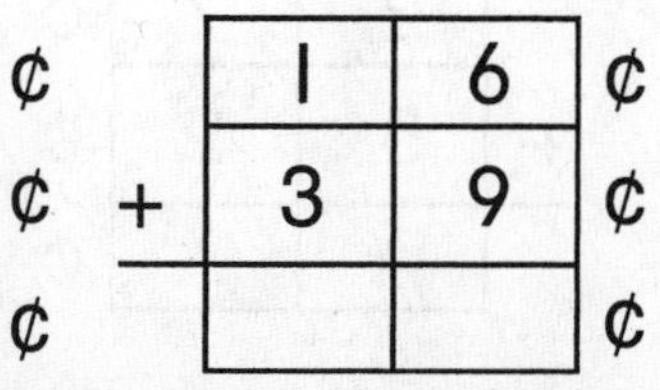

Name ______________________

Date ______________________

Homework 64B

***Saxon** Math 2 (for use with **Lesson 64**)*

1. Melvin wears a different pair of socks each day.
Draw a picture to show the pairs of socks he wore last week.

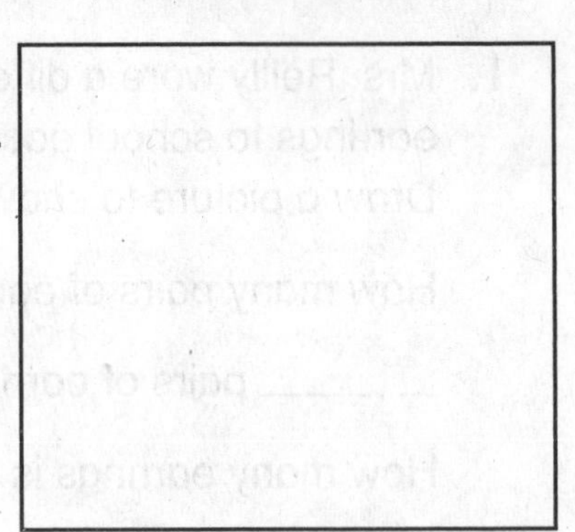

(Remember, there are ________ days in one week.)

How many pairs of socks is that?

________ pairs of socks

How many socks is that? ________ socks

2. Draw a line of symmetry in each shape.
Color the shape with 4 angles green.
Color the shape with 5 angles yellow.
Color the shape with 3 angles orange.

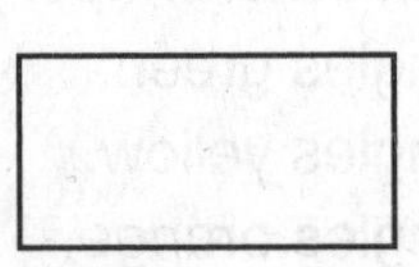
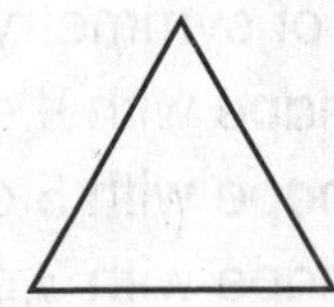
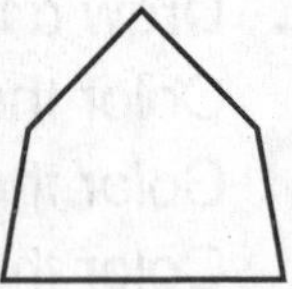

3. Draw 6 pairs of gloves. How many gloves is that? ________

4. Two of these quadrilaterals are congruent (same shape and same size).
Color the congruent quadrilaterals red.

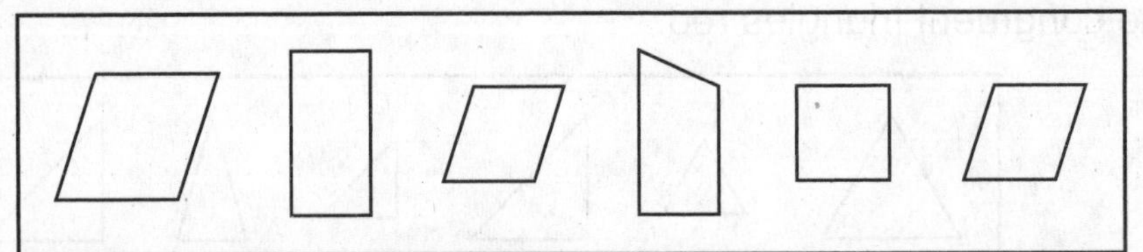

5. About how long is your bed?

4 inches   6 feet   10 inches   2 feet

6. Find the sums.

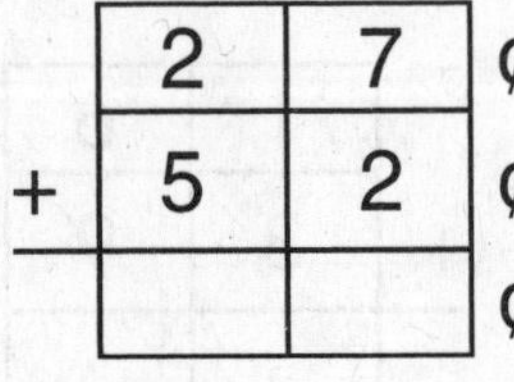

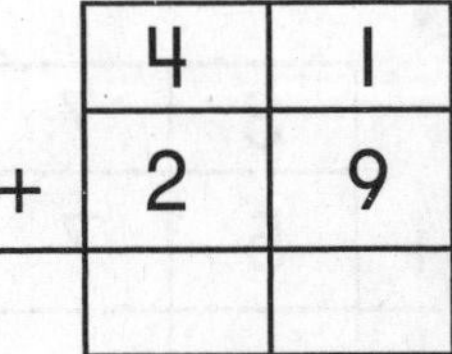

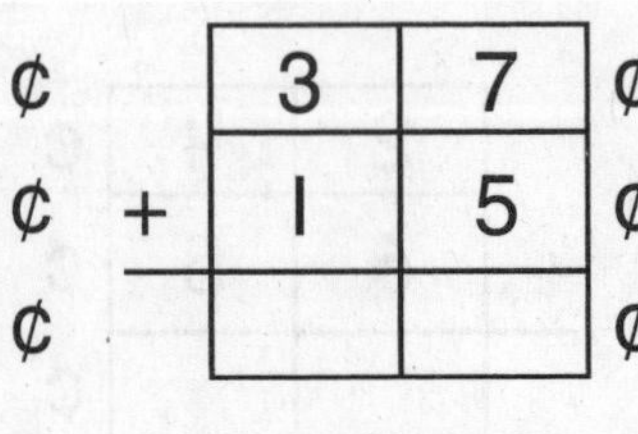

	5	9	¢
+	2	3	¢
			¢

Name ______________________ Score ______

Fact Assessment 12

Saxon Math 2 (for use with Lesson 65-1)

Set 12: 100 Addition Facts

	9	2	6	5	0	9	7	1	2	5
1	+ 1	+ 2	+ 4	+ 1	+ 7	+ 9	+ 3	+ 6	+ 5	+ 4
	9	2	8	4	6	7	3	9	0	4
2	+ 4	+ 0	+ 7	+ 1	+ 6	+ 8	+ 2	+ 8	+ 8	+ 6
	5	3	0	8	3	7	7	1	6	2
3	+ 2	+ 9	+ 6	+ 1	+ 3	+ 4	+ 0	+ 5	+ 7	+ 3
	1	5	7	3	2	9	7	4	0	6
4	+ 0	+ 5	+ 6	+ 4	+ 1	+ 5	+ 2	+ 9	+ 3	+ 8
	8	3	1	0	6	5	1	8	2	5
5	+ 2	+ 5	+ 7	+ 0	+ 2	+ 7	+ 4	+ 6	+ 9	+ 0
	6	0	3	4	9	1	6	2	8	0
6	+ 3	+ 5	+ 7	+ 4	+ 2	+ 8	+ 5	+ 4	+ 8	+ 9
	4	7	9	9	5	0	3	7	6	4
7	+ 2	+ 7	+ 0	+ 6	+ 8	+ 1	+ 6	+ 9	+ 0	+ 8
	7	2	4	1	4	8	3	8	1	5
8	+ 1	+ 6	+ 7	+ 2	+ 5	+ 9	+ 0	+ 3	+ 9	+ 6
	1	3	0	5	9	2	8	4	6	1
9	+ 1	+ 8	+ 2	+ 9	+ 3	+ 7	+ 0	+ 3	+ 9	+ 3
	8	4	5	2	3	7	9	0	8	6
10	+ 5	+ 0	+ 3	+ 8	+ 1	+ 5	+ 7	+ 4	+ 4	+ 1

Name ____________________

Date ____________________

Guided Class Practice 65A

***Saxon** Math 2 (for use with **Lesson 65-1**)*

1. David has 7 dimes and 3 pennies.
John has 5 pennies and 2 dimes.
How much money do the two boys have altogether?

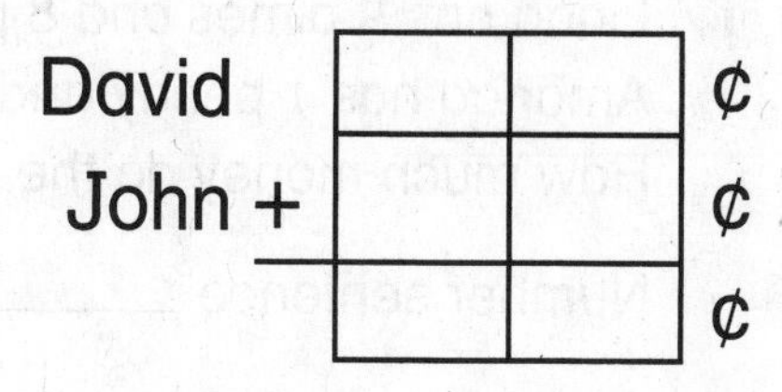

Number sentence ____________________

Answer ____________________

2. What coins could you use to make **27¢?**

____

3. Draw a shape that has 4 angles in the box.

How many sides does the shape have? ______

4. Write the fraction that tells how much is shaded.

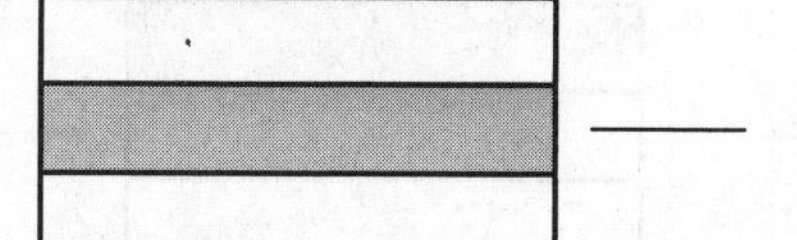 ____ 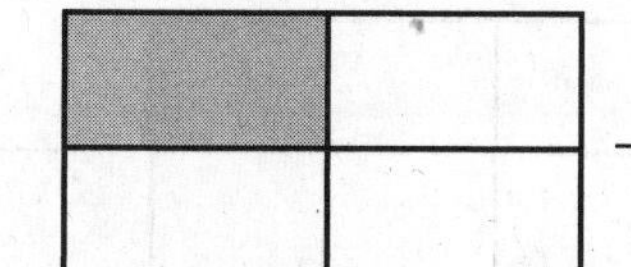 ____ 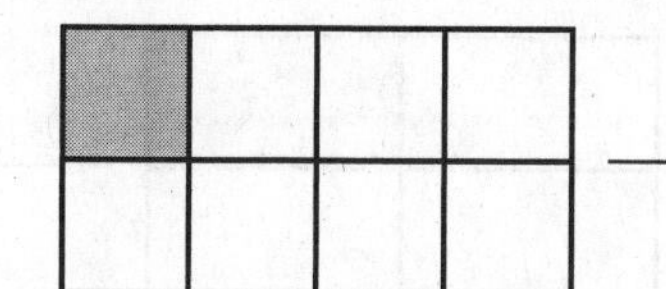 ____

5. Write the problems vertically. Find the sums.

82¢ + 16¢

17¢ + 48¢

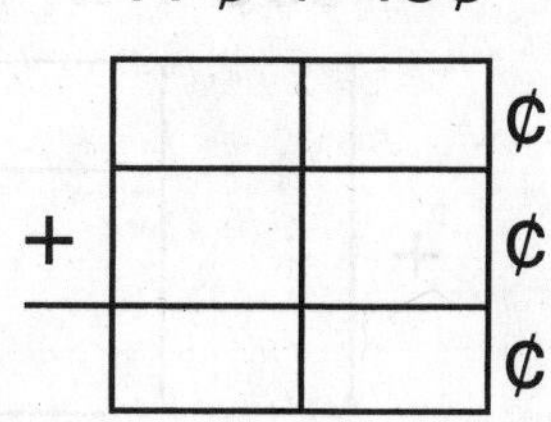

44¢ + 36¢

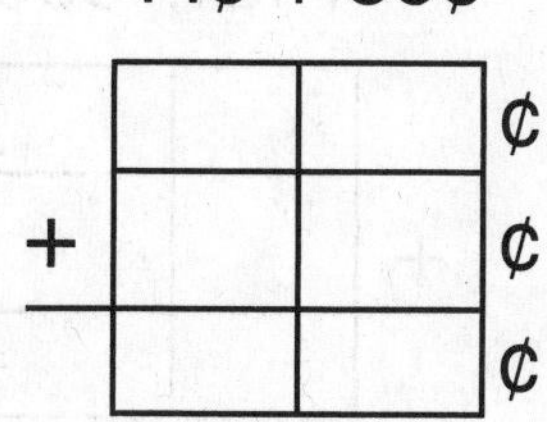

74¢ + 9¢

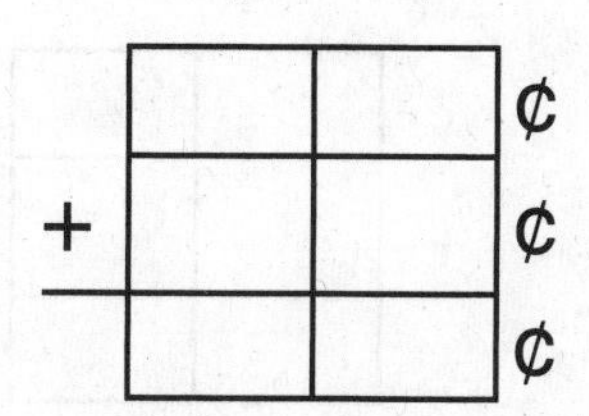

6. Find the sums.

6 + 2 + 3 + 8 = ______

5 + 4 + 5 + 3 + 7 + 2 = ______

Name ____________________

Date ____________________

Homework 65B

Saxon Math 2 (for use with **Lesson 65-1**)

1. Linda has 4 dimes and 8 pennies.
Amanda has 1 penny and 3 dimes.
How much money do the two girls have altogether?

Linda ¢
Amanda + ¢
¢

Number sentence ____________________

Answer ____________________

2. What coins could you use to make **23¢?**

____

3. Draw a shape that has 3 angles in the box.

How many sides does the shape have? ______

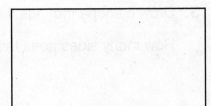

4. Write the fraction that tells how much is shaded.

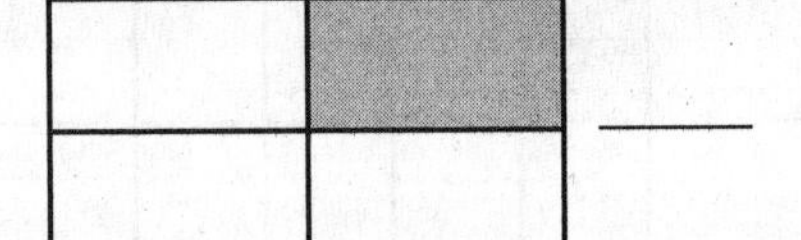 ______ 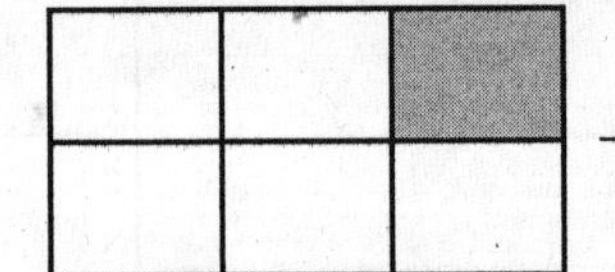 ______ 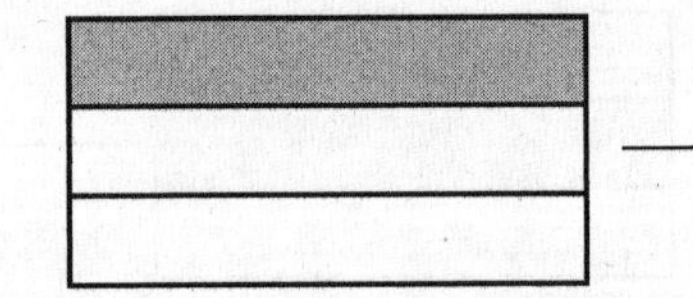 ______

5. Write the problems vertically. Find the sums.

51¢ + 16¢

14¢ + 36¢ 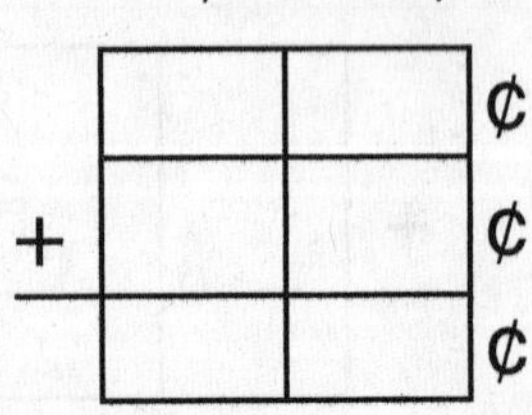

55¢ + 29¢ 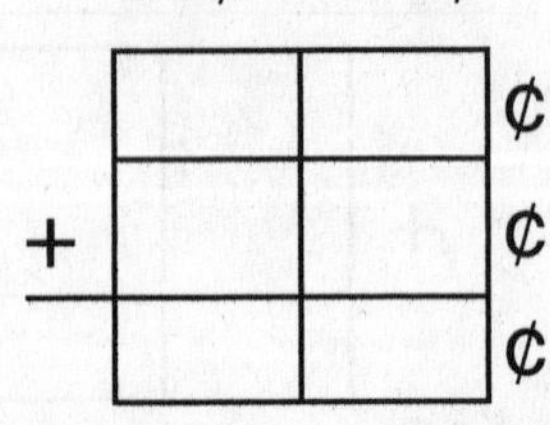

7¢ + 37¢ 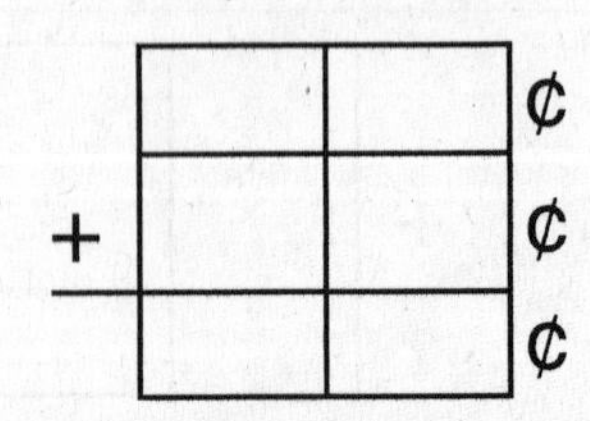

6. Find the sums.

$2 + 3 + 5 + 7 =$ ______

$6 + 2 + 6 + 3 + 4 + 8 =$ ______

Name ______________________

Date ______________________

Saxon *Math 2 (for use with* ***Lesson 65-2****)*

1. Daniel had four nickels. Roseann gave him five more nickels. How many nickels does he have now?

Number sentence ______________________

Answer ______________________

2. Finish numbering the number line.

Put a point at **3.** Label it **A.**
Put a point at **6.** Label it **B.**

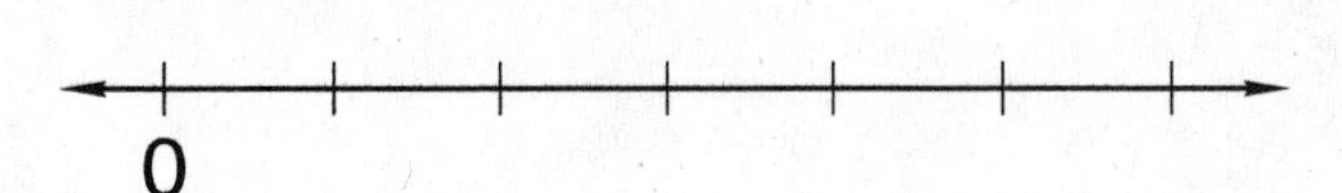

3. How much money is this? __________

4. Darlene has 4 pairs of mittens.
Draw the mittens.
How many mittens did you draw? ________

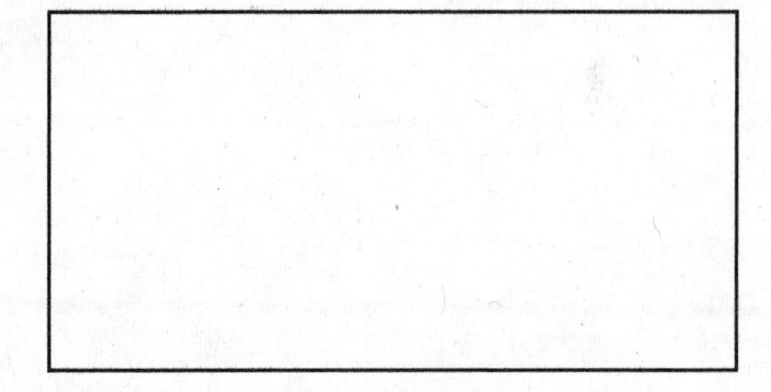

5. Draw a line of symmetry in each shape or letter.

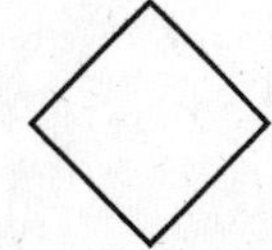

6. Find the sums.

58 + 10 = ________

10 + 40 = ________

	dimes	pennies	
	2	8	¢
+	3	1	¢
			¢

	dimes	pennies	
	1	4	¢
+	3	3	¢
			¢

Name ________________________________

Lesson Worksheet 65-2

***Saxon** Math 2 (for use with **Lesson 65-2**)*

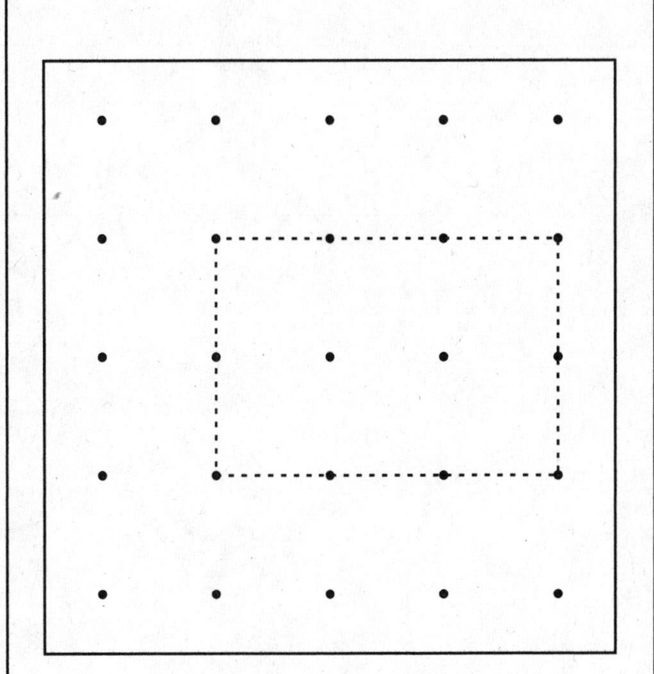

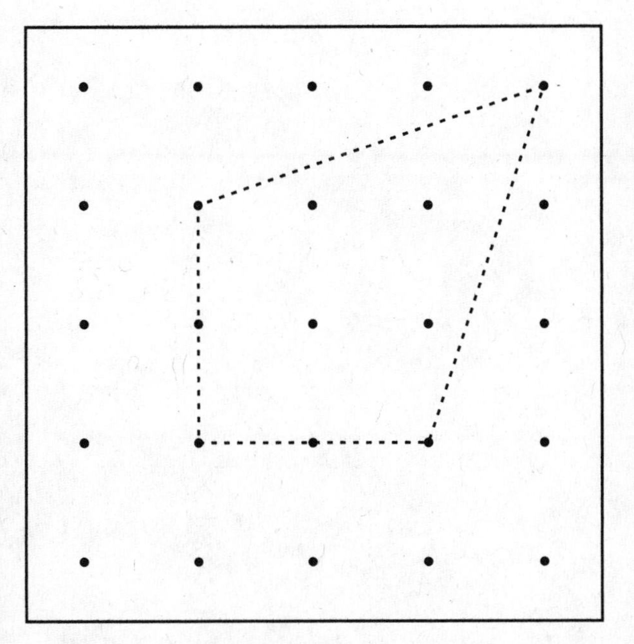

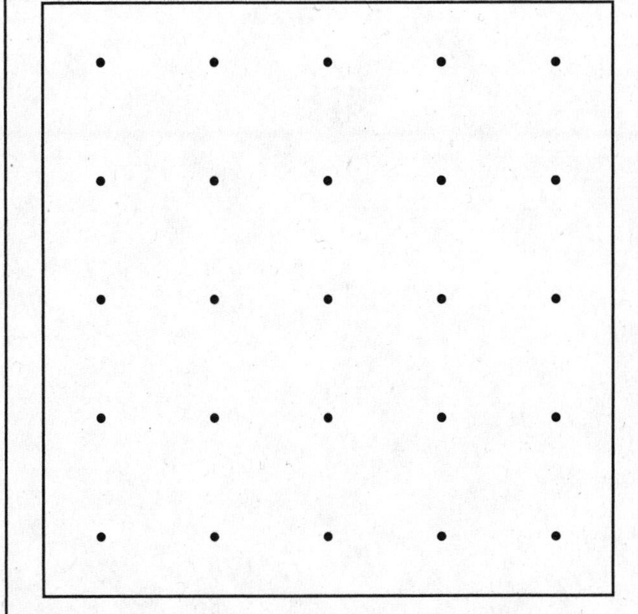

Name ______________________________

Class Fact Practice 66A

***Saxon** Math 2 (for use with **Lesson 66**)*

Set 13: Subtracting 1 and 0

Do −1 Wrap-Up once.
Do −0 Wrap-Up once.

2 − 1 = ___	7 − 1 = ___	10 − 1 = ___	5 − 1 = ___	3 − 1 = ___
8 − 1 = ___	4 − 1 = ___	1 − 1 = ___	9 − 1 = ___	6 − 1 = ___
6 − 0 = ___	3 − 0 = ___	9 − 0 = ___	7 − 0 = ___	0 − 0 = ___
5 − 0 = ___	2 − 0 = ___	8 − 0 = ___	1 − 0 = ___	4 − 0 = ___
10 − 1 = ___	5 − 0 = ___	3 − 1 = ___	6 − 0 = ___	7 − 1 = ___

Name ______________________________

Fact Homework 66B

***Saxon** Math 2 (for use with **Lesson 66**)*

Set 13: Subtracting 1 and 0

1. Read the answers to someone.
2. Write the answers.
3. Ask someone to correct your paper. Corrected by ____________________

3 − 1	7 − 1	1 − 1	9 − 1	6 − 1
2 − 1	10 − 1	5 − 1	8 − 1	4 − 1
0 − 0	8 − 0	3 − 0	6 − 0	2 − 0
4 − 0	7 − 0	1 − 0	9 − 0	5 − 0
6 − 1	2 − 0	9 − 1	8 − 0	4 − 1

Name ______________________

Date ______________________

Guided Class Practice 66A

Saxon Math 2 (for use with Lesson 66)

1. Joe had 4 dimes. His mother gave him 4 more dimes. Write a number sentence to show how many dimes he has now.

Number sentence ______________________

Answer ______________

How much money is that? __________

2. Use the graph to answer the questions.

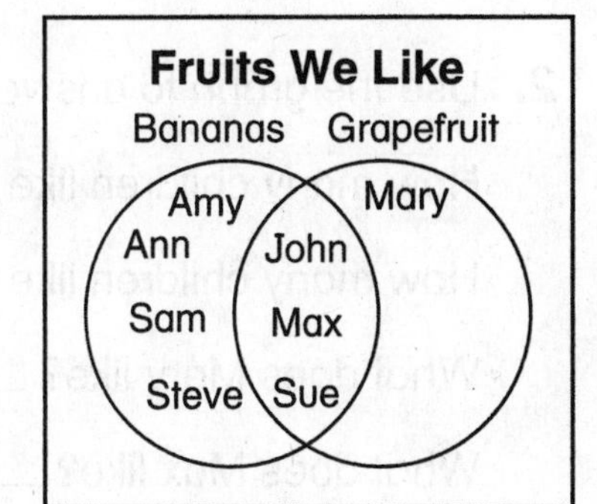

How many children like bananas? ________

How many children like only grapefruit? ________

How many children like both bananas and grapefruit? ________

What does Amy like? ______________

3. Measure each line segment using inches.

vertical line segment ________"

horizontal line segment ________"

oblique line segment ________"

4. Write the fraction that tells what fractional part of the set of flowers is shaded. ——

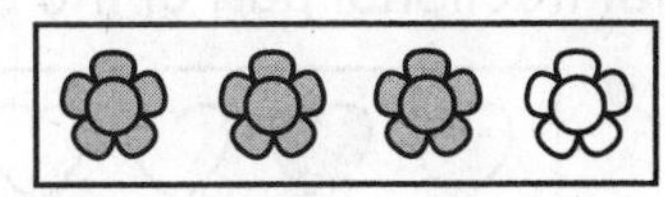

5. Find the sums.

21¢ + 48¢

+ ¢ ¢ ¢

35¢ + 55¢

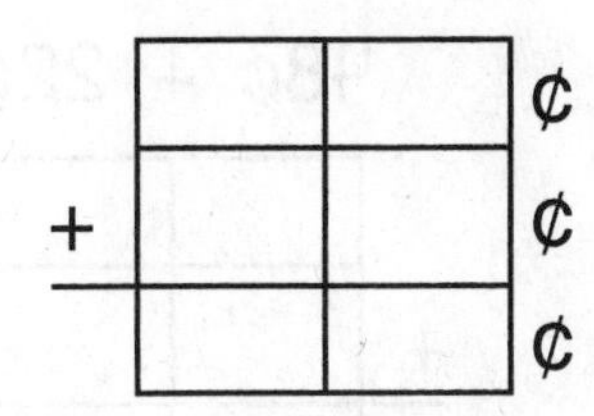

46¢ + 27¢

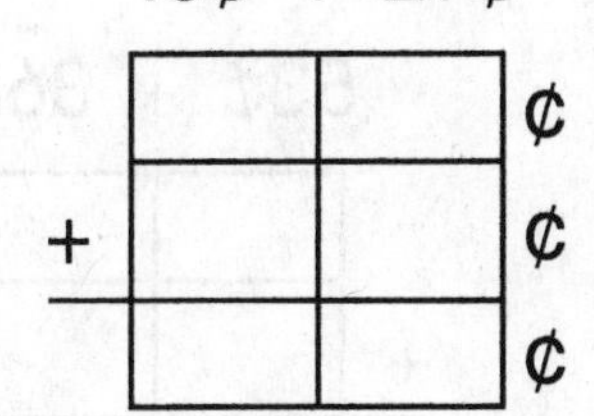

Name ______________________

Date ______________________

Homework 66B

Saxon Math 2 (for use with Lesson 66)

1. Micky had 3 dimes. His brother gave him 2 more dimes. Write a number sentence to show how many dimes he has now.

Number sentence ______________________

Answer ______________________

How much money is that? __________

2. Use the graph to answer the questions.

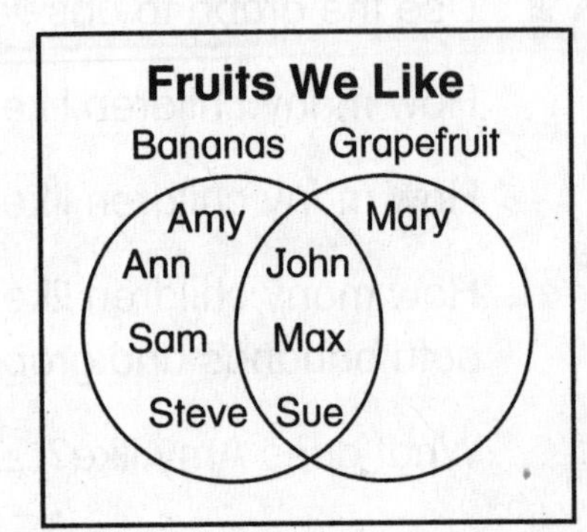

How many children like grapefruit? ______

How many children like only bananas? ______

What does Mary like? ______________

What does Max like? ______________

____

3. Finish these number patterns.

66, 67, 68, ____, ____, ____, ____, ____, ____

5, 10, 15, ____, ____, ____, ____, ____, ____, ____

4. Write the fraction that tells what fractional part of the set of hearts is shaded. ____

5. Find the sums.

53¢ + 36¢

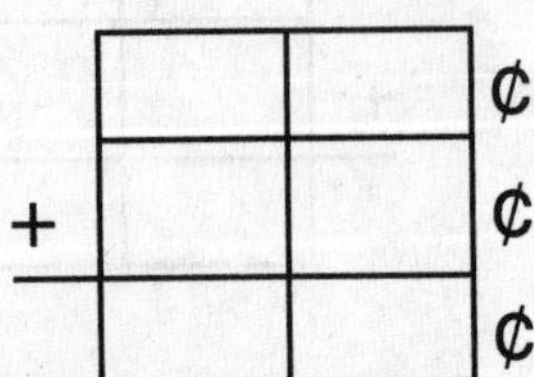

48¢ + 22¢

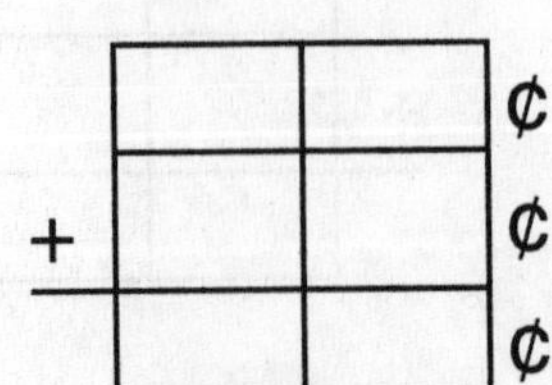

29¢ + 17¢

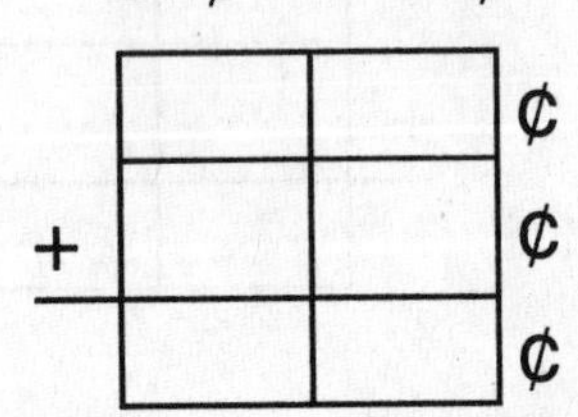

Name ____________________

Set 13: Subtracting 1 and 0

Do −1 Wrap-Up once.
Do −0 Wrap-Up once.

$4 - 0$	$7 - 1$	$3 - 1$	$1 - 0$	$6 - 1$
$8 - 0$	$5 - 1$	$3 - 0$	$10 - 1$	$4 - 1$
$0 - 0$	$1 - 1$	$3 - 1$	$6 - 0$	$9 - 1$
$5 - 1$	$9 - 0$	$8 - 1$	$2 - 1$	$2 - 0$
$7 - 0$	$9 - 1$	$5 - 0$	$7 - 1$	$1 - 1$

Name ______________________________

Fact Homework 67B

Saxon *Math 2 (for use with* ***Lesson 67****)*

Set 13: Subtracting 1 and 0

1. Read the answers to someone.
2. Write the answers.
3. Ask someone to correct your paper. Corrected by ______________________

$$\begin{array}{r} 1 \\ -\ 0 \\ \hline \end{array} \quad \begin{array}{r} 7 \\ -\ 1 \\ \hline \end{array} \quad \begin{array}{r} 6 \\ -\ 0 \\ \hline \end{array} \quad \begin{array}{r} 5 \\ -\ 1 \\ \hline \end{array} \quad \begin{array}{r} 4 \\ -\ 1 \\ \hline \end{array}$$

$$\begin{array}{r} 1 \\ -\ 1 \\ \hline \end{array} \quad \begin{array}{r} 8 \\ -\ 0 \\ \hline \end{array} \quad \begin{array}{r} 3 \\ -\ 1 \\ \hline \end{array} \quad \begin{array}{r} 9 \\ -\ 1 \\ \hline \end{array} \quad \begin{array}{r} 0 \\ -\ 0 \\ \hline \end{array}$$

$$\begin{array}{r} 10 \\ -\ 1 \\ \hline \end{array} \quad \begin{array}{r} 2 \\ -\ 0 \\ \hline \end{array} \quad \begin{array}{r} 8 \\ -\ 1 \\ \hline \end{array} \quad \begin{array}{r} 4 \\ -\ 0 \\ \hline \end{array} \quad \begin{array}{r} 2 \\ -\ 1 \\ \hline \end{array}$$

$$\begin{array}{r} 7 \\ -\ 1 \\ \hline \end{array} \quad \begin{array}{r} 5 \\ -\ 0 \\ \hline \end{array} \quad \begin{array}{r} 6 \\ -\ 1 \\ \hline \end{array} \quad \begin{array}{r} 1 \\ -\ 1 \\ \hline \end{array} \quad \begin{array}{r} 9 \\ -\ 0 \\ \hline \end{array}$$

$$\begin{array}{r} 3 \\ -\ 0 \\ \hline \end{array} \quad \begin{array}{r} 5 \\ -\ 1 \\ \hline \end{array} \quad \begin{array}{r} 7 \\ -\ 0 \\ \hline \end{array} \quad \begin{array}{r} 3 \\ -\ 1 \\ \hline \end{array} \quad \begin{array}{r} 9 \\ -\ 1 \\ \hline \end{array}$$

Name ____________________

Date ____________________

Guided Class Practice 67A

***Saxon** Math 2 (for use with **Lesson 67**)*

1. John bought a dozen ice cream bars. He ate one ice cream bar. How many ice cream bars are left?

Number sentence ____________________

Answer ____________________

2. Use the graph to answer the questions.

How many children have a dog? ______

How many children have only cats? ______

How many children have both a cat and a dog? ______

What pet does Mike have? ______

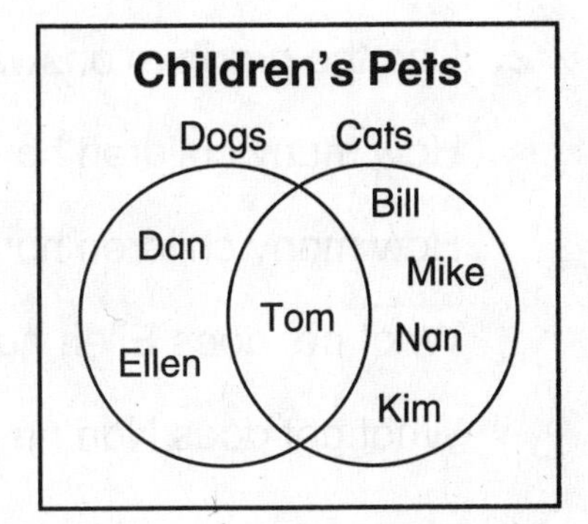

3. It is morning.
What time is it?

______

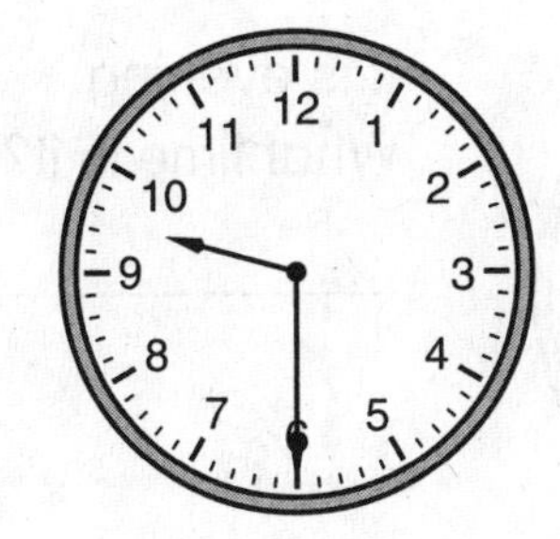

It is evening.
What time is it?

______

4. Write the fraction that tells how much is shaded.

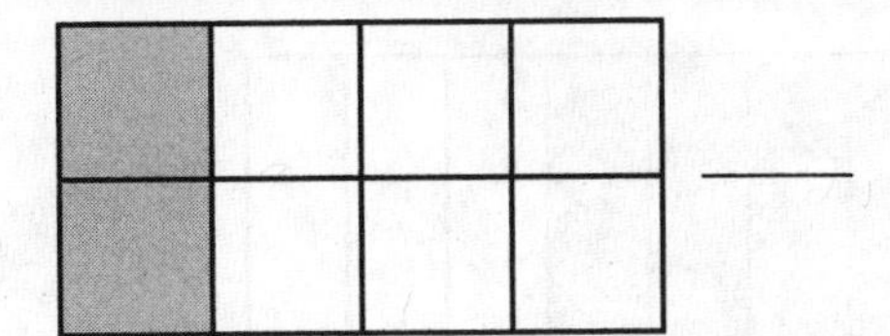

___

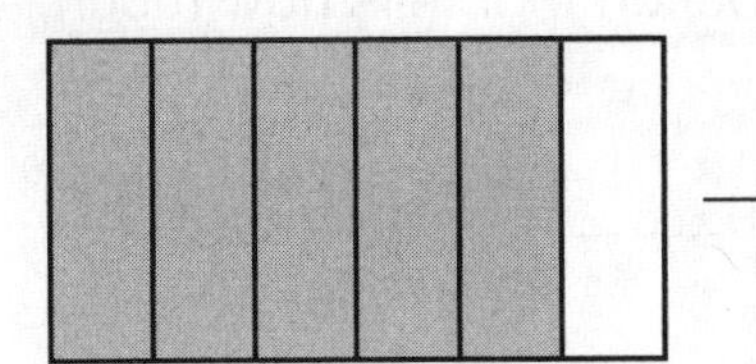

___

5. Find the sums.

$$\begin{array}{r} 6 \\ 2 \\ 3 \\ +\ 4 \\ \hline \end{array}$$

$$\begin{array}{rrrl} & 5 & 1 & ¢ \\ + & 2 & 8 & ¢ \\ \hline & & & ¢ \end{array}$$

$$\begin{array}{rrrl} & 3 & 5 & ¢ \\ + & 1 & 6 & ¢ \\ \hline & & & ¢ \end{array}$$

$$\begin{array}{rrrl} & 4 & 7 & ¢ \\ + & 4 & 7 & ¢ \\ \hline & & & ¢ \end{array}$$

Name ____________________

Date ____________________

Homework 67B

***Saxon** Math 2 (for use with **Lesson 67**)*

1. Mr. Brandon bought a dozen doughnuts. He ate two on the way home. How many doughnuts are left?

Number sentence ____________________

Answer ____________________

2. Use the graph to answer the questions.

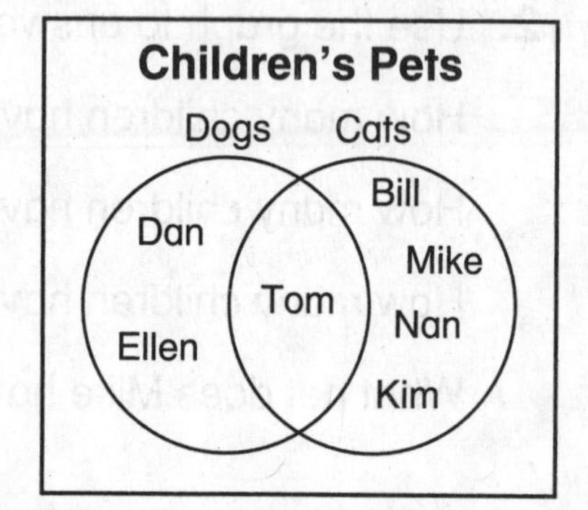

How many children have a cat? ________

How many children have only dogs? ________

What pet does Ellen have? ________

What pet does Nan have? ________

3. It is morning.
What time is it?

____

It is evening.
What time is it?

____

4. Write the fraction that tells how much is shaded.

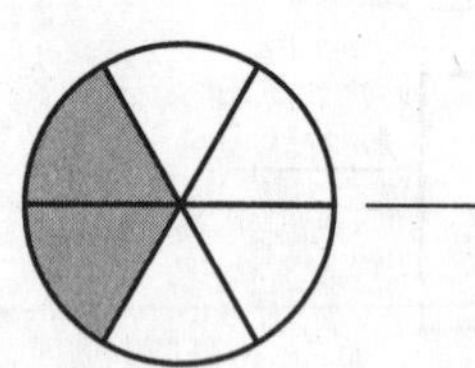 ____

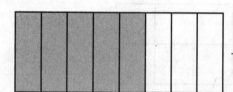 ____

5. Find the sums.

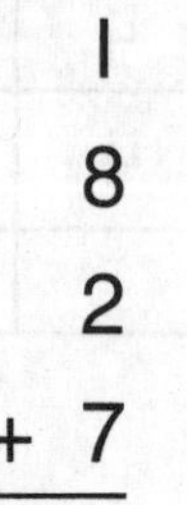

$$\begin{array}{r} 1 \\ 8 \\ 2 \\ +\ 7 \\ \hline \end{array}$$

	3	2	¢
+	1	7	¢
			¢

	2	8	¢
+	2	8	¢
			¢

		9	¢
+	4	5	¢
			¢

Name ______________________

Class Fact Practice 68A

***Saxon** Math 2 (for use with **Lesson 68**)*

Set 13: Subtracting 1 and 0

Do −1 Wrap-Up twice.
Do −0 Wrap-Up once.

$8 - 1$	$9 - 0$	$4 - 1$	$1 - 1$	$6 - 0$
$7 - 1$	$4 - 0$	$10 - 1$	$0 - 0$	$5 - 1$
$9 - 1$	$7 - 0$	$3 - 1$	$6 - 1$	$1 - 1$
$3 - 0$	$5 - 1$	$1 - 0$	$9 - 1$	$5 - 0$
$2 - 1$	$8 - 0$	$7 - 1$	$2 - 0$	$3 - 1$

Name ______________________________

Fact Homework 68B

***Saxon** Math 2 (for use with **Lesson 68**)*

Set 13: Subtracting 1 and 0

1. Read the answers to someone.
2. Write the answers.
3. Ask someone to correct your paper. Corrected by ______________________

$1 - 1$	$7 - 1$	$5 - 0$	$9 - 1$	$7 - 0$
$2 - 0$	$2 - 1$	$8 - 1$	$9 - 0$	$5 - 1$
$9 - 1$	$6 - 0$	$3 - 1$	$1 - 1$	$0 - 0$
$4 - 1$	$10 - 1$	$3 - 0$	$5 - 1$	$8 - 0$
$6 - 1$	$1 - 0$	$3 - 1$	$7 - 1$	$4 - 0$

Name ______________________________

***Saxon** Math 2 (for use with **Lesson 68**)*

1.

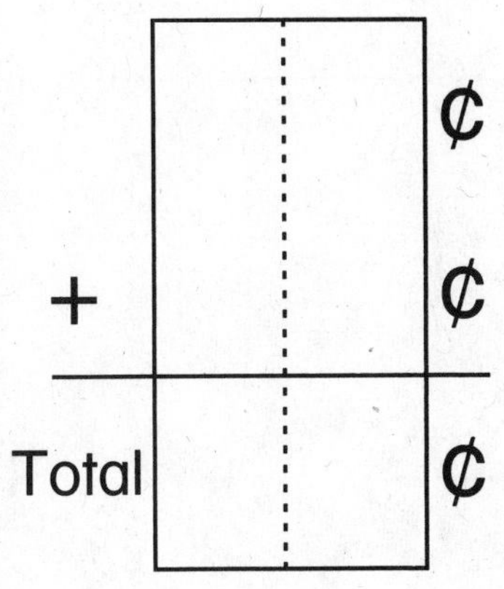

2.

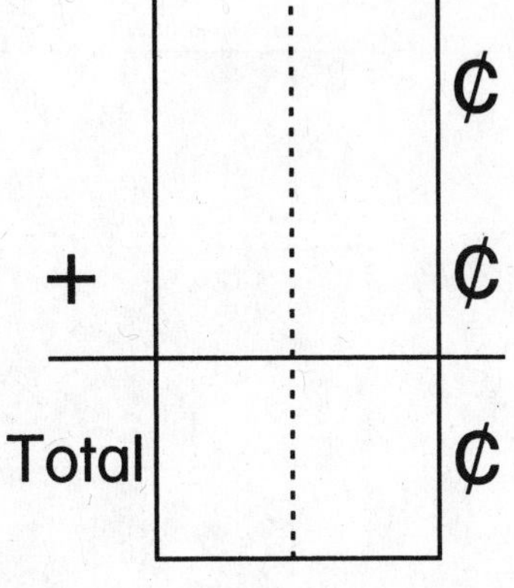

3.

¢

\+ ¢

Total ¢

4.

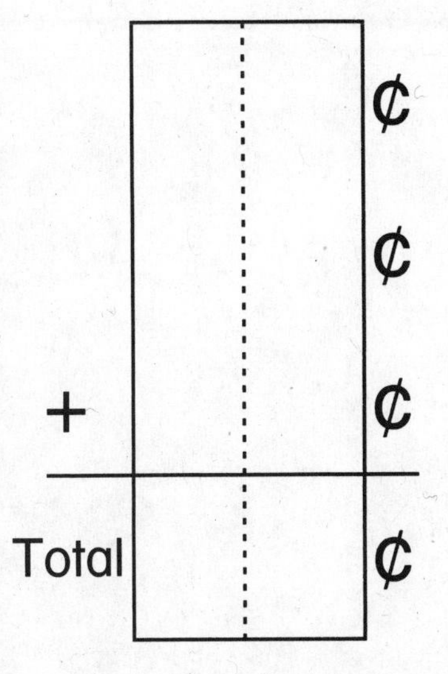

5.

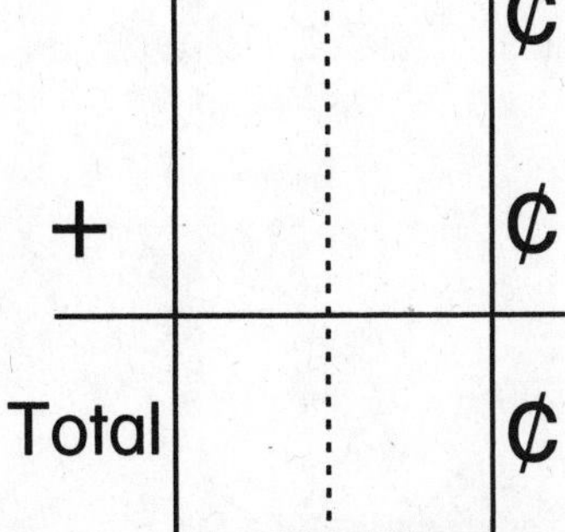

6.

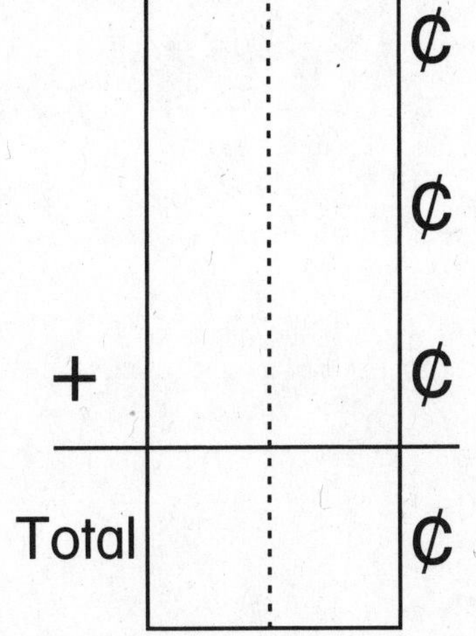

Name ____________________

Date ____________________

Guided Class Practice 68A

***Saxon** Math 2 (for use with **Lesson 68**)*

1. Four children voted for Apple A. Seven children voted for Apple C, and three children voted for Apple B. Color the graph to show how many children voted for each apple. How many children voted altogether?

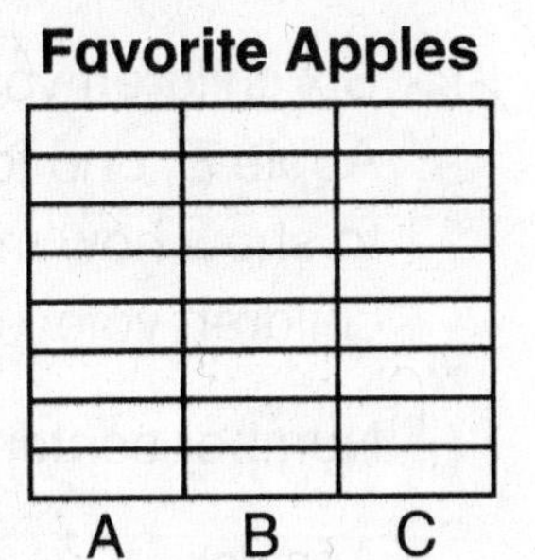

Number sentence ____________________

Answer ____________________

2. Write these numbers in order from least to greatest.

49	25	63	28	42

_____ _____ _____ _____ _____

least ... greatest

3. Use the graph to answer the questions.

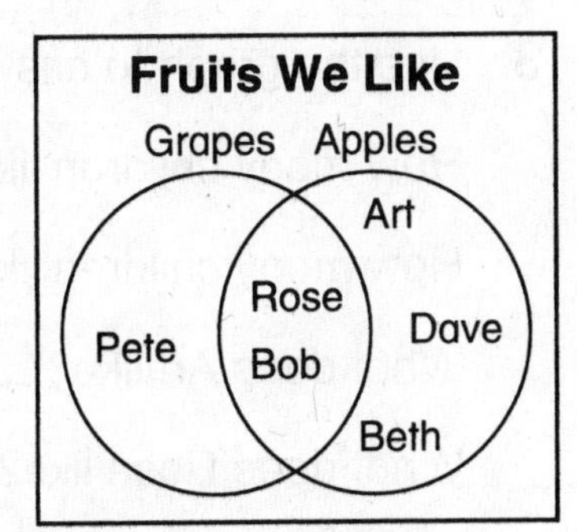

How many children like apples? _____

How many children like only grapes? _____

How many children like both grapes and apples? _____

What does Rose like? ____________________

4. Cheryl has a dozen apples and a half-dozen oranges. Draw the fruit.

5. It's half past three in the morning. Write the digital time.

[:]

○ a.m.

○ p.m.

6. Find the sums.

	1	6	¢
	2	2	¢
+	4	8	¢
			¢

	2	7	¢
	3	1	¢
+	1	3	¢
			¢

Name ____________________

Date ____________________

Homework 68B

***Saxon** Math 2 (for use with **Lesson 68**)*

1. Six children voted for Apple D. Three children voted for Apple E, and four children voted for Apple F. Color the graph to show how many children voted for each apple. How many children voted altogether?

Number sentence ____________________

Answer ____________________

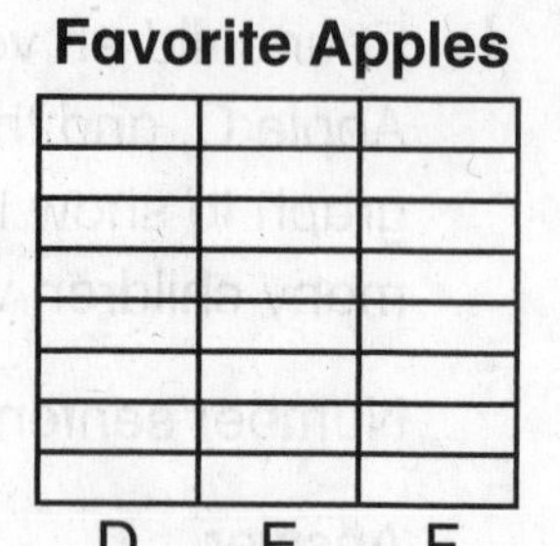

2. Write these numbers in order from least to greatest.

53	71	37	58	33

____ ____ ____ ____ ____
least ... greatest

3. Use the graph to answer the questions.

How many children like grapes? ______

How many children like only apples? ______

What does Art like? ____________________

What does Rose like? ____________________

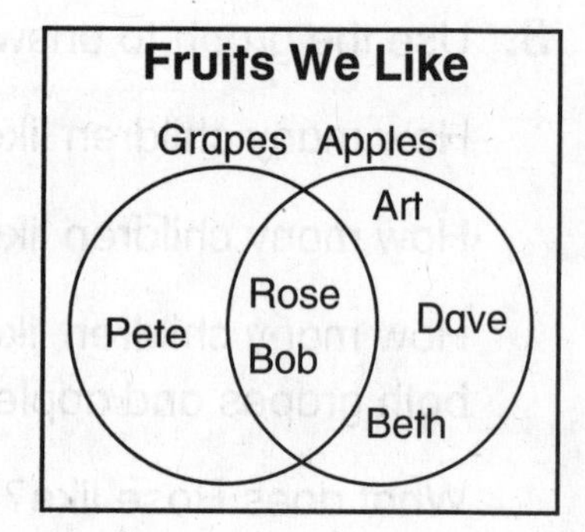

4. Curtis has a half-dozen peaches and a dozen bananas. Draw the fruit.

5. It's half past four in the afternoon. Write the digital time.

____ : ____ ○ a.m. ○ p.m.

6. Find the sums.

	1	3	¢
	4	2	¢
+	4	1	¢
			¢

	2	4	¢
	1	7	¢
+	3	3	¢
			¢

Name ____________________

Class Fact Practice 69A

***Saxon** Math 2 (for use with **Lesson 69**)*

Set 13: Subtracting 1 and 0

Do −1 Wrap-Up once.
Do −0 Wrap-Up once.

$$\begin{array}{r}9\\-\ 1\\\hline\end{array}\quad\begin{array}{r}3\\-\ 1\\\hline\end{array}\quad\begin{array}{r}7\\-\ 0\\\hline\end{array}\quad\begin{array}{r}5\\-\ 1\\\hline\end{array}\quad\begin{array}{r}3\\-\ 0\\\hline\end{array}$$

$$\begin{array}{r}9\\-\ 0\\\hline\end{array}\quad\begin{array}{r}1\\-\ 1\\\hline\end{array}\quad\begin{array}{r}6\\-\ 1\\\hline\end{array}\quad\begin{array}{r}5\\-\ 0\\\hline\end{array}\quad\begin{array}{r}7\\-\ 1\\\hline\end{array}$$

$$\begin{array}{r}2\\-\ 1\\\hline\end{array}\quad\begin{array}{r}4\\-\ 0\\\hline\end{array}\quad\begin{array}{r}8\\-\ 1\\\hline\end{array}\quad\begin{array}{r}2\\-\ 0\\\hline\end{array}\quad\begin{array}{r}10\\-\ 1\\\hline\end{array}$$

$$\begin{array}{r}0\\-\ 0\\\hline\end{array}\quad\begin{array}{r}9\\-\ 1\\\hline\end{array}\quad\begin{array}{r}3\\-\ 1\\\hline\end{array}\quad\begin{array}{r}8\\-\ 0\\\hline\end{array}\quad\begin{array}{r}1\\-\ 1\\\hline\end{array}$$

$$\begin{array}{r}4\\-\ 1\\\hline\end{array}\quad\begin{array}{r}5\\-\ 1\\\hline\end{array}\quad\begin{array}{r}6\\-\ 0\\\hline\end{array}\quad\begin{array}{r}7\\-\ 1\\\hline\end{array}\quad\begin{array}{r}1\\-\ 0\\\hline\end{array}$$

Name ______________________________

Set 13: Subtracting 1 and 0

1. Read the answers to someone.
2. Write the answers.
3. Ask someone to correct your paper. Corrected by ____________________

3 − 1	2 − 0	7 − 1	8 − 0	2 − 1
5 − 0	9 − 1	1 − 0	5 − 1	3 − 0
1 − 1	6 − 1	3 − 1	7 − 0	9 − 1
5 − 1	0 − 0	10 − 1	4 − 0	7 − 1
6 − 0	1 − 1	4 − 1	9 − 0	8 − 1

Name ________________________________

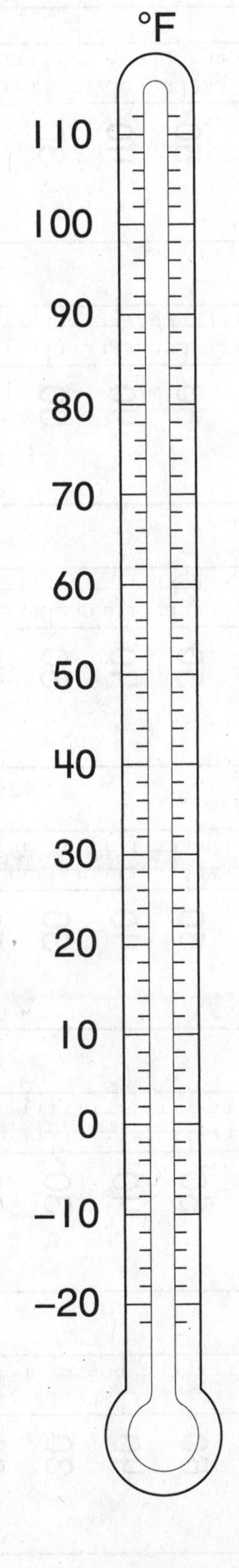

Name ____________________

1.

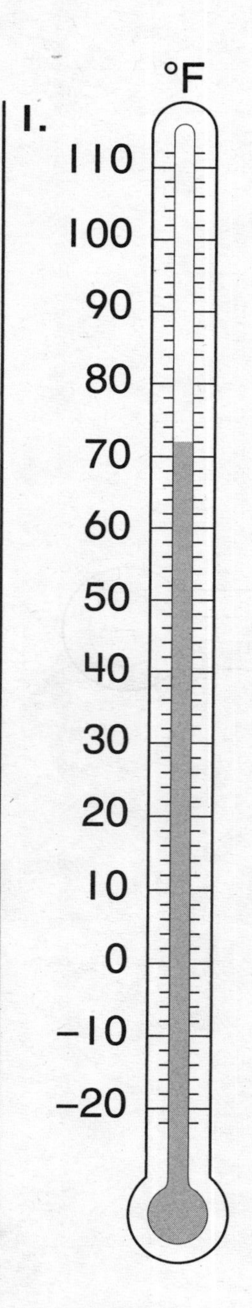

2.

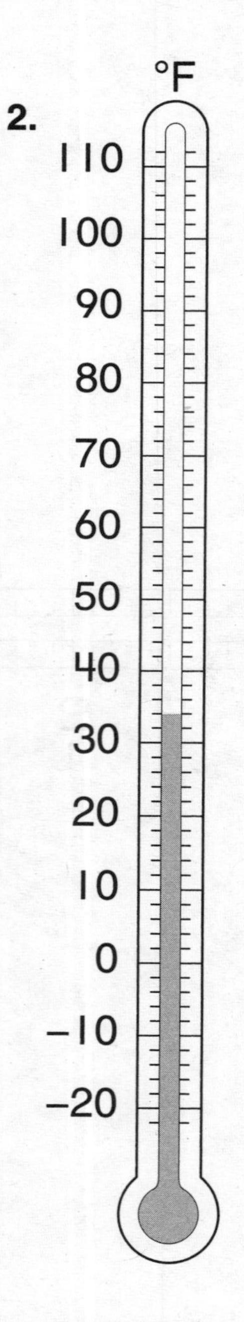

3.

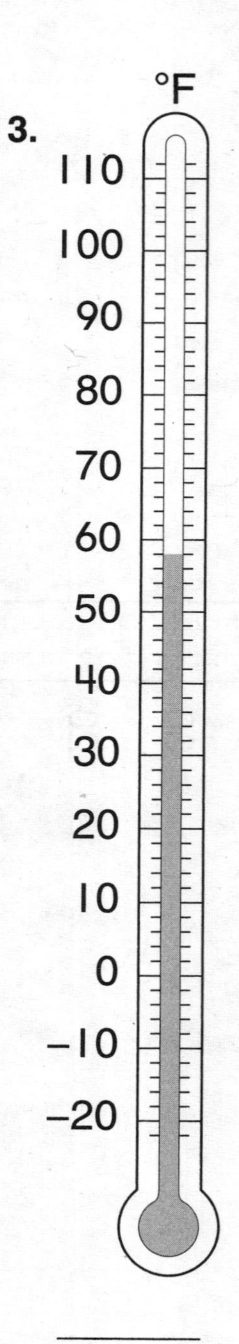

4.

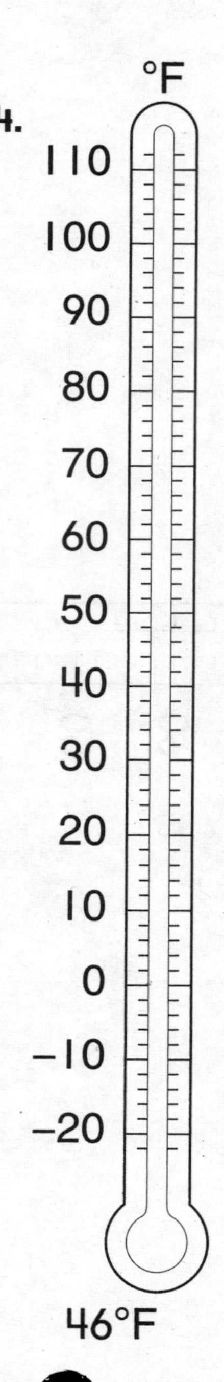

46°F

5.

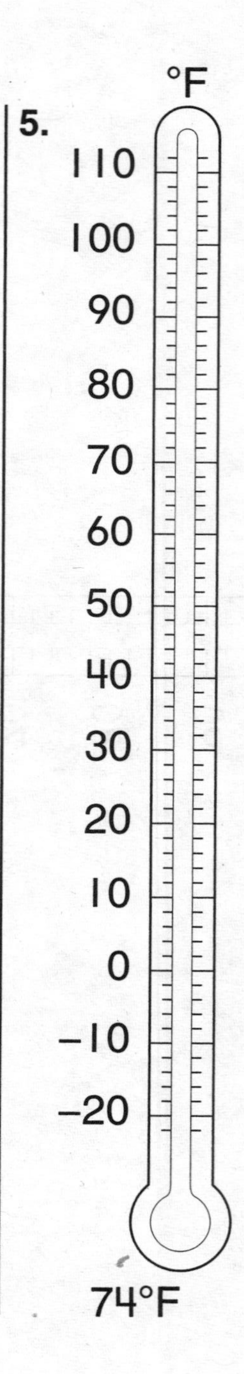

74°F

6.

28°F

Name ______________________

Date ______________________

Workspace

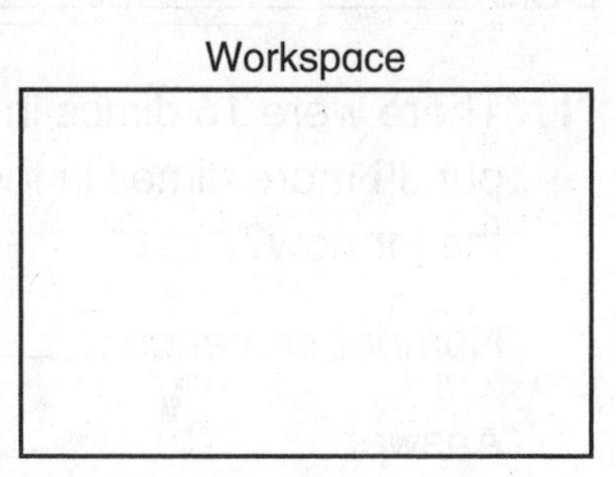

1. There were 43 pennies in the penny jar. Mrs. Kaplan put 17 more pennies in the jar. How many pennies are in the jar now?

   Number sentence ______________________

   Answer ______________________

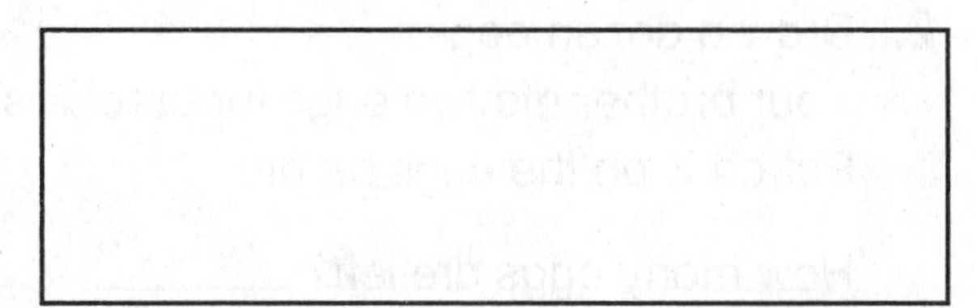

2. Draw one dozen doughnuts.
   Your sister ate half a dozen doughnuts.
   Put an X on the doughnuts she ate.

   How many doughnuts are left? ________

3. Use a red crayon to color all the rectangles that are congruent to the rectangle on the left. (Congruent means same shape and same size.)

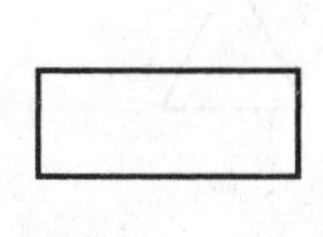

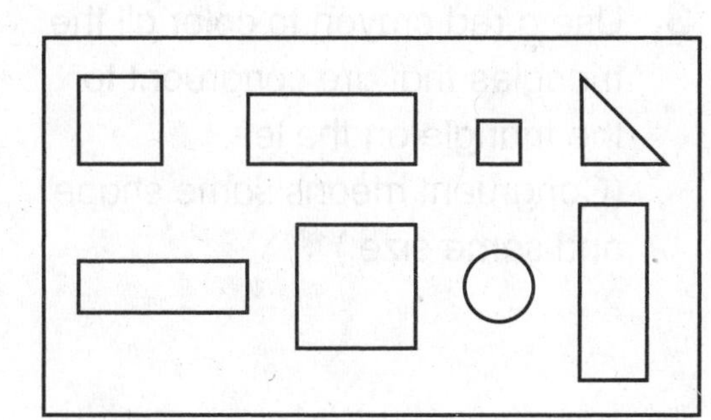

4. What is the temperature on the thermometer? ________°F

5. It's half past eight in the morning.

   Is it a.m. or p.m.? ____________

   Show the time on both clocks.

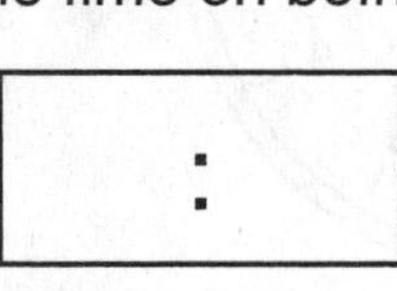

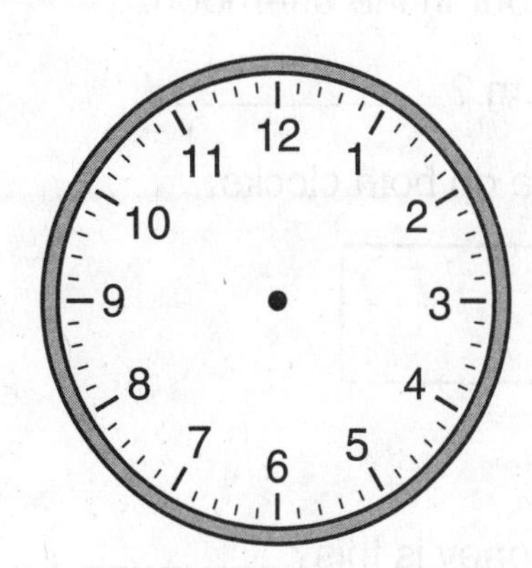

°F

110
100
90
80
70
60
50
40
30
20
10
0
−10
−20

6. How much money is this? ____________

Name ______________________________

Date ______________________________

Saxon Math 2 (for use with Lesson 69)

Workspace

1. There were 16 dimes in the dime jar. Mrs. McDonough put 34 more dimes in the jar. How many dimes are in the jar now?

Number sentence ______________________

Answer ______________________

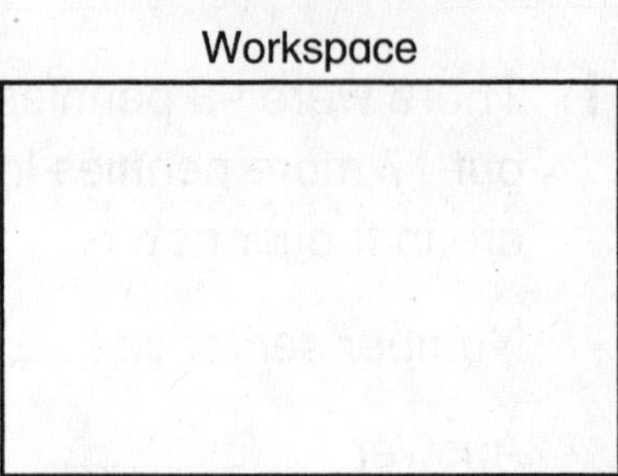

2. Draw a dozen eggs.
Your brother ate two eggs for breakfast.
Put an X on the eggs he ate.

How many eggs are left? ________

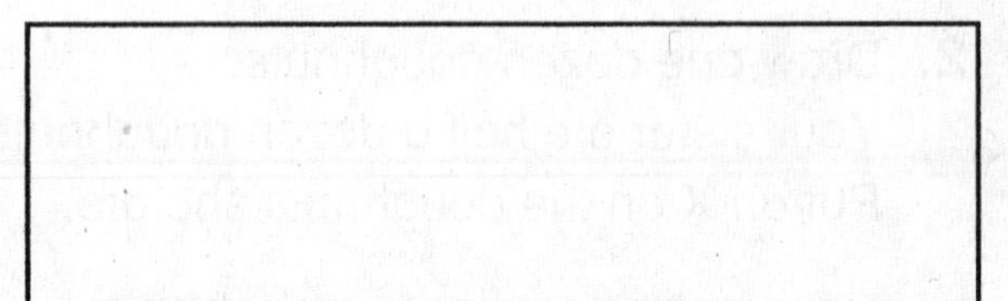

3. Use a red crayon to color all the triangles that are congruent to the triangle on the left.
(Congruent means same shape and same size.)

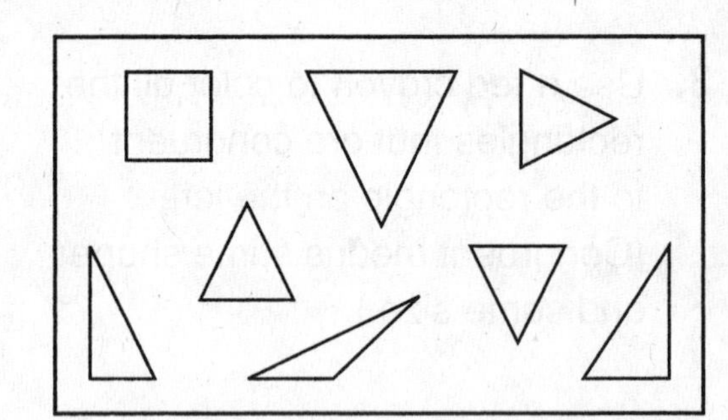

4. What is the temperature on the thermometer? ________°F

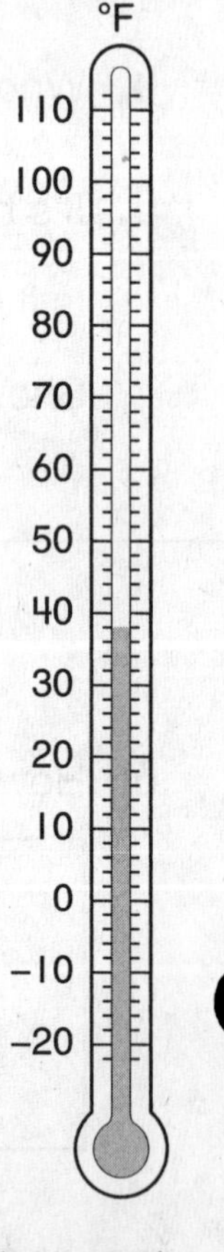

5. It's half past four in the afternoon.

Is it a.m. or p.m.? ____________

Show the time on both clocks.

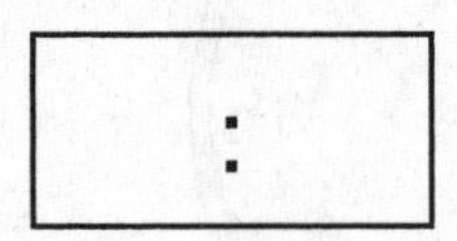

6. How much money is this? ____________

Name ______________________ Score ______

Fact Assessment 13-1

***Saxon** Math 2 (for use with **Lesson 70-1**)*

Set 13: Subtracting 1 and 0

5 − 1	4 − 1	9 − 0	1 − 1	7 − 1
10 − 1	7 − 0	6 − 1	3 − 0	1 − 1
3 − 1	9 − 1	2 − 0	8 − 1	5 − 0
7 − 1	4 − 0	2 − 1	0 − 0	5 − 1
9 − 1	6 − 0	1 − 0	3 − 1	8 − 0

Name ______________________________

Fact Homework 70A

***Saxon** Math 2 (for use with **Lesson 70-1**)*

Set 13: Subtracting 1 and 0

Pretend you are the teacher.
Correct this paper.
If the answer is incorrect, write the correct answer next to the problem.

5	4	9	1	7
− 1	− 1	− 0	− 1	− 1
4	3	9	1	6

10	7	6	3	1
− 1	− 0	− 1	− 0	− 1
9	7	7	3	0

3	9	2	8	5
− 1	− 1	− 0	− 1	− 0
2	8	2	7	5

7	4	2	0	5
− 1	− 0	− 1	− 0	− 1
5	4	1	0	4

9	6	1	3	8
− 1	− 0	− 0	− 1	− 0
8	6	1	3	8

Name ____________________

Fact Homework 70B

***Saxon** Math 2 (for use with **Lesson 70-1**)*

A100: 100 Addition Facts Corrected by ________

1	9 + 1	2 + 2	6 + 4	5 + 1	0 + 7	9 + 9	7 + 3	1 + 6	2 + 5	5 + 4
2	9 + 4	2 + 0	8 + 7	4 + 1	6 + 6	7 + 8	3 + 2	9 + 8	0 + 8	4 + 6
3	5 + 2	3 + 9	0 + 6	8 + 1	3 + 3	7 + 4	7 + 0	1 + 5	6 + 7	2 + 3
4	1 + 0	5 + 5	7 + 6	3 + 4	2 + 1	9 + 5	7 + 2	4 + 9	0 + 3	6 + 8
5	8 + 2	3 + 5	1 + 7	0 + 0	6 + 2	5 + 7	1 + 4	8 + 6	2 + 9	5 + 0
6	6 + 3	0 + 5	3 + 7	4 + 4	9 + 2	1 + 8	6 + 5	2 + 4	8 + 8	0 + 9
7	4 + 2	7 + 7	9 + 0	9 + 6	5 + 8	0 + 1	3 + 6	7 + 9	6 + 0	4 + 8
8	7 + 1	2 + 6	4 + 7	1 + 2	4 + 5	8 + 9	3 + 0	8 + 3	1 + 9	5 + 6
9	1 + 1	3 + 8	0 + 2	5 + 9	9 + 3	2 + 7	8 + 0	4 + 3	6 + 9	1 + 3
10	8 + 5	4 + 0	5 + 3	2 + 8	3 + 1	7 + 5	9 + 7	0 + 4	8 + 4	6 + 1

Name ______________________________

A. Write the answers.

2 – 2 = ______

3 – 2 = ______

4 – 2 = ______

5 – 2 = ______

6 – 2 = ______

7 – 2 = ______

8 – 2 = ______

9 – 2 = ______

10 – 2 = ______

11 – 2 = ______

12 – 2 = ______

13 – 2 = ______

B. Draw lines to connect the problems to the answers.

Name ______________________

Lesson Worksheet 70-1B

Saxon *Math 2 (for use with* ***Lesson 70-1****)*

Do –2 Wrap-Up once. ☐ Do –2 Wrap-Up once. ☐ Do –2 Wrap-Up once. ☐

A.	**B.**	**C.**
$5 - 2 =$ ____	$7 - 2 =$ ____	$9 - 2 =$ ____
$9 - 2 =$ ____	$2 - 2 =$ ____	$3 - 2 =$ ____
$11 - 2 =$ ____	$10 - 2 =$ ____	$11 - 2 =$ ____
$4 - 2 =$ ____	$6 - 2 =$ ____	$7 - 2 =$ ____
$8 - 2 =$ ____	$13 - 2 =$ ____	$2 - 2 =$ ____
$10 - 2 =$ ____	$8 - 2 =$ ____	$13 - 2 =$ ____
$3 - 2 =$ ____	$3 - 2 =$ ____	$8 - 2 =$ ____
$12 - 2 =$ ____	$11 - 2 =$ ____	$4 - 2 =$ ____
$7 - 2 =$ ____	$4 - 2 =$ ____	$6 - 2 =$ ____
$2 - 2 =$ ____	$12 - 2 =$ ____	$12 - 2 =$ ____
$13 - 2 =$ ____	$5 - 2 =$ ____	$10 - 2 =$ ____
$6 - 2 =$ ____	$9 - 2 =$ ____	$5 - 2 =$ ____

Name ____________________

Date ____________________

Problem-Solving Worksheet 70A

***Saxon** Math 2 (for use with **Lesson 70-1**)*

Understand	Plan	Solve	Check

Guess and Check

At the M. G. Ellis School Store children can buy pencils and erasers. Pencils cost 3¢ and erasers cost 2¢. Nikita bought pencils and erasers. She spent 10¢. Show what she bought.

How many pencils and erasers did Nikita buy?

________ pencils ________ erasers

Name ______________________

Understand	Plan	Solve	Check

At the Harbor school store children can buy stickers. Happy-face stickers cost 1¢ each and star stickers cost 3¢ each. William bought some happy-face stickers and some star stickers. He spent 9¢ for stickers. Show which stickers he bought.

How many happy-face stickers and star stickers did William buy?

_______ happy-face and _______ star or

_______ happy-face and _______ star

Circle the problem-solving strategies you used to solve this problem.

Act It Out ***Use Logical Reasoning***

Draw a Picture ***Look for a Pattern***

Make an Organized List ***Guess and Check***

Explain how you got your answer: ______________________

____

____

____

Name ____________________ Score ________

Fact Assessment 13-2

***Saxon** Math 2 (for use with **Lesson 70-2**)*

A100: 100 Addition Facts

1	9 + 1	2 + 2	6 + 4	5 + 1	0 + 7	9 + 9	7 + 3	1 + 6	2 + 5	5 + 4
2	9 + 4	2 + 0	8 + 7	4 + 1	6 + 6	7 + 8	3 + 2	9 + 8	0 + 8	4 + 6
3	5 + 2	3 + 9	0 + 6	8 + 1	3 + 3	7 + 4	7 + 0	1 + 5	6 + 7	2 + 3
4	1 + 0	5 + 5	7 + 6	3 + 4	2 + 1	9 + 5	7 + 2	4 + 9	0 + 3	6 + 8
5	8 + 2	3 + 5	1 + 7	0 + 0	6 + 2	5 + 7	1 + 4	8 + 6	2 + 9	5 + 0
6	6 + 3	0 + 5	3 + 7	4 + 4	9 + 2	1 + 8	6 + 5	2 + 4	8 + 8	0 + 9
7	4 + 2	7 + 7	9 + 0	9 + 6	5 + 8	0 + 1	3 + 6	7 + 9	6 + 0	4 + 8
8	7 + 1	2 + 6	4 + 7	1 + 2	4 + 5	8 + 9	3 + 0	8 + 3	1 + 9	5 + 6
9	1 + 1	3 + 8	0 + 2	5 + 9	9 + 3	2 + 7	8 + 0	4 + 3	6 + 9	1 + 3
10	8 + 5	4 + 0	5 + 3	2 + 8	3 + 1	7 + 5	9 + 7	0 + 4	8 + 4	6 + 1

Name ______________________________

Date ______________________________

1. Phil had 7 dimes. His brother gave him 2 dimes. How many dimes does Phil have now?

Number sentence ______________________________

Answer ____________________

How much money is this? __________

2. Write these numbers in order from least to greatest.

48	27	25	43	39

_____ _____ _____ _____ _____

least — greatest

3. Write the fractions that show what part is shaded.

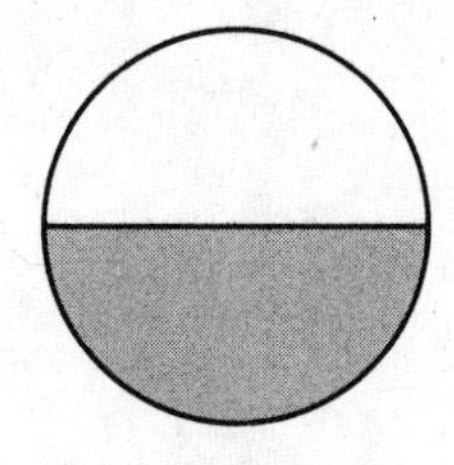 ____

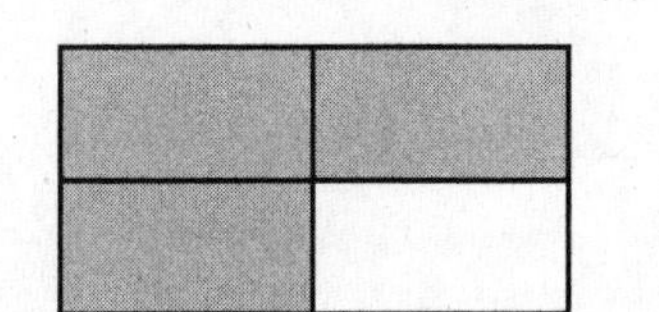 ____

4. Draw a 4-inch line segment.

•

5. Put a dot inside each angle.
Count the number of angles in each shape.

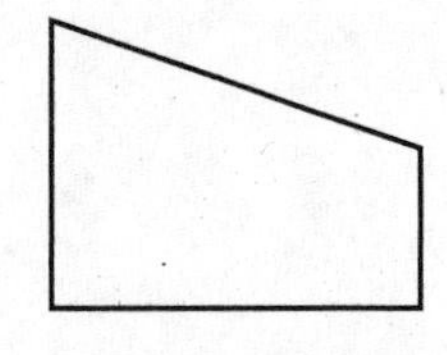

_______ angles

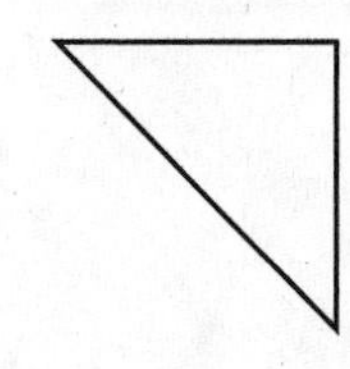

_______ angles

6. Add.

	1	6	¢
+	7	2	¢
			¢

	2	4	¢
+	5	3	¢
			¢

$$\begin{array}{r} 4 \\ 2 \\ 1 \\ +\ 8 \\ \hline \end{array}$$

$6 + 4 + 7 + 2 + 3 =$ _______

Name ______________________________

How many squares or triangles can you find in each design?

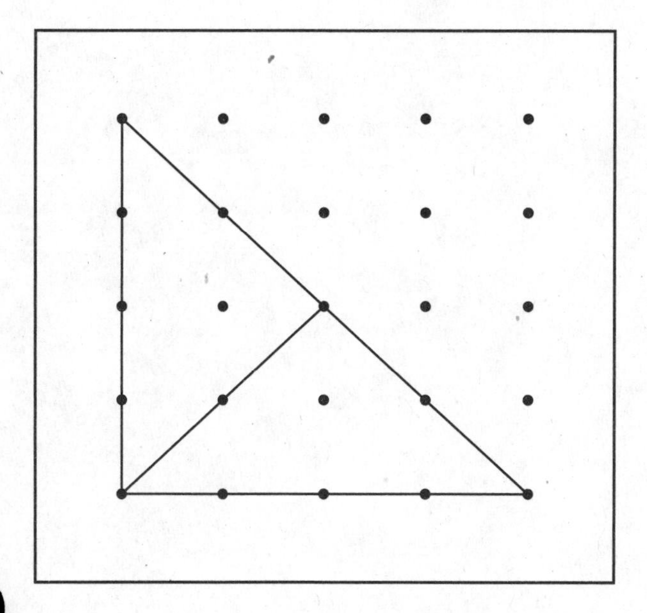

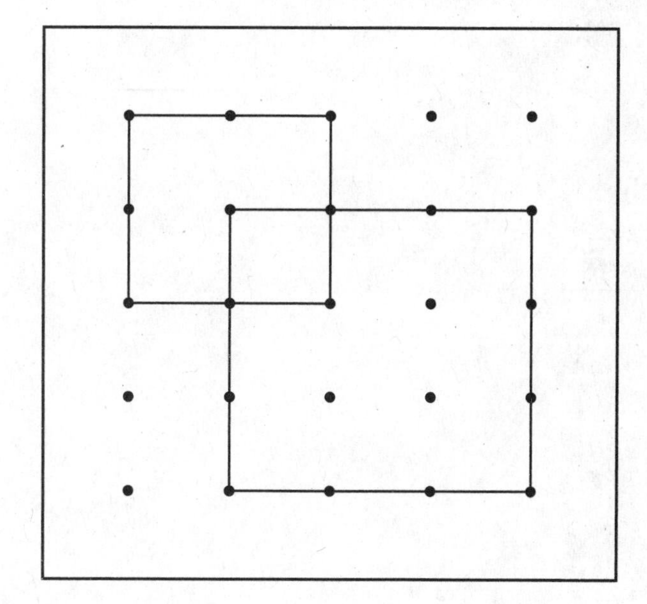

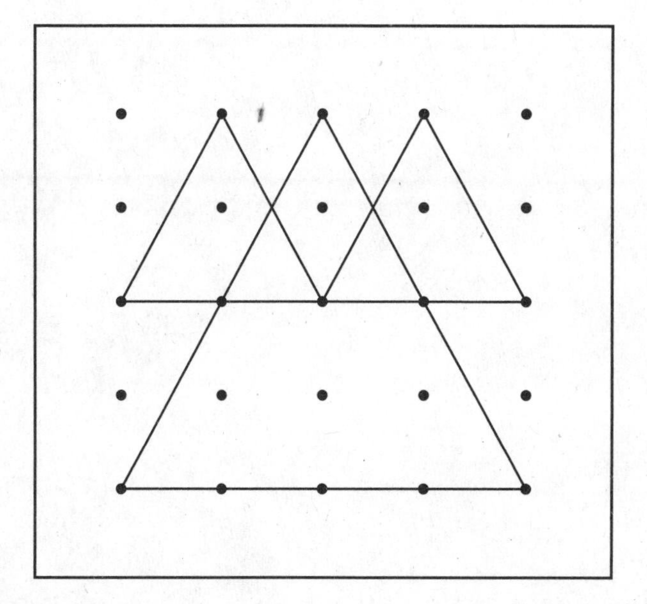

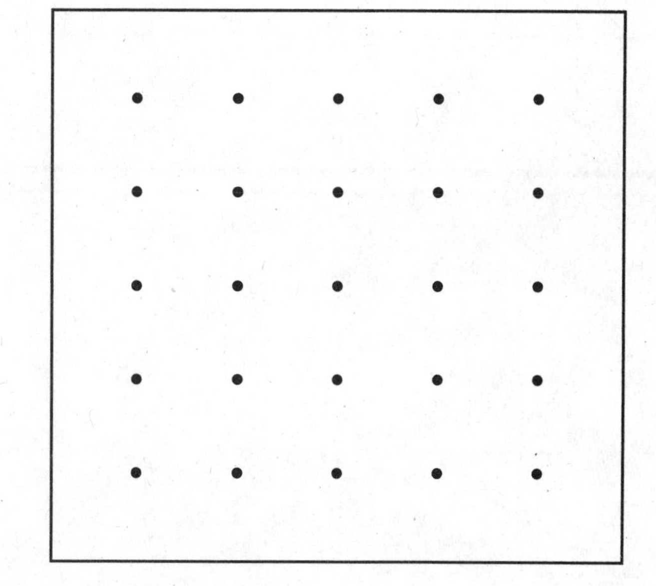